Reprint Publishing

FOR PEOPLE WHO GO FOR ORIGINALS.

www.reprintpublishing.com

LECTURES ON BACTERIA

DE BARY

𝕷𝖔𝖓𝖉𝖔𝖓

HENRY FROWDE

OXFORD UNIVERSITY PRESS WAREHOUSE

AMEN CORNER, E.C.

LECTURES

ON

BACTERIA

BY

A. DE BARY

PROFESSOR IN THE UNIVERSITY OF STRASSBURG

SECOND IMPROVED EDITION

AUTHORISED TRANSLATION BY

HENRY E. F. GARNSEY, M.A.

Fellow of Magdalen College, Oxford

REVISED BY

ISAAC BAYLEY BALFOUR, M.A., M.D., F.R.S.

*Fellow of Magdalen College and Sherardian Professor of Botany
in the University of Oxford*

WITH 20 WOOD-ENGRAVINGS

Oxford

AT THE CLARENDON PRESS

1887

PREFACE TO THE ENGLISH EDITION.

THIS translation of Professor De Bary's 'Vorlesungen über Bacterien' has been prepared because there is at present no book in English which gives in like manner 'a general view of the subject' of Bacteria, and 'sets forth the known facts in the life of Bacteria in their connection with those with which we are acquainted in other branches of natural history.'

I. B. B.

OXFORD, 1887.

AUTHOR'S PREFACE.

THE present work is in the main a short abridgement of a number of lectures, some of which were delivered in a connected series as a University course, others as occasional and separate addresses. The form of the lectures has been occasionally altered to meet the difference between a written treatise and free oral delivery accompanied by demonstrations. Some things have been omitted and others added, especially some matters of general importance which were not published or did not become known to me till after the delivery of the actual course.

The lectures were an attempt to introduce an audience composed of persons of very different professional pursuits, medical and non-medical, to an acquaintance with the present state of knowledge and opinion concerning the much discussed questions connected with Bacteria. They had, therefore, to give such a survey of the subject as would be intelligible to all who were not strangers to the elements of a scientific training, and especially to set forth the known facts in the life of the Bacteria in their connection with those with which we are acquainted in other branches of natural history.

A survey of the present extensive literature of the subject, and of the almost daily additions to it, shows the existence of many serviceable and some excellent publications, but at the same time also of much that is mistaken and obscure. The scientific and semi-scientific converse of the day, if I may use the expression, is greatly influenced by works of the latter kind, and the chief reason for this, if I am not mistaken, lies in the absence of a general view of the subject itself and of its relations to other portions of natural history; we cannot see the wood for the trees. An attempt to give such a view would be no mere superfluous addition to existing works, and this consideration was a decisive reason in the judgment of myself and of those who gave me their encouragement for afterwards transcribing and publishing my lectures.

The present treatise, therefore, must not be expected to be a Bacteriology, or even to report and enumerate all the details which may be of interest and importance ; it should rather serve only as a guide for the direction of the student through these details.

Many readers, devoted to the study of the Bacteria, will be familiar with the literature or with the guides to it before they take up this book. For the sake of those who seek to gain some knowledge of the subject from its perusal, and also for the purpose of naming the most important sources of information which I have made use of along with my own investigations, I have added a few notices of publications at the end of the volume, and have indicated by numerals in brackets the places in the text to which the citation marked with the same number refers.

So much by way of introduction to this little work. I trust that it may do something to clear up existing views on the subject of the Bacteria, and to lead the investigation of these organisms from its present stage of storm and pressure into the ways of quiet fruitful labour and increase of knowledge.

The above with the omission of one sentence is the wording of the preface written in July, 1885, for the first edition of this book. The kindly reception which it met with can only have been due to the circumstance that the form in which the subject was presented in it was the one best adapted to attain the object proposed; it could scarcely be that there was anything new in it. Hence the form and limits of the second edition which is now demanded are alike prescribed to me; it must be made as like as possible to the first. This has been done; the original frame is unaltered, and the old matter still appears in it in many places. On the other hand, much progress has been made in the period which has elapsed since the work was originally composed, and some new views have been laid down which could not be disregarded. The new edition, therefore, will be found to contain not only some editorial improvements in the special descriptions, but also various important alterations.

These observations apply also to the notes at the end of the work, except that I have introduced somewhat more critical and explanatory remarks than in the first edition.

A. DE BARY.

STRASSBURG, *Oct.*, 1886.

CONTENTS.

Contents.

I.

Introduction. Bacteria or Schizomycetes and Fungi. Structure of the Bacterium-cell (1).

THE purpose of these lectures is to give some account of the present state of our knowledge respecting the objects included under the name of Bacteria. It is unnecessary to enlarge upon the manifold interest attaching to these organisms at a time when the statement urged daily on the educated public does not fall far short of saying, that a large part of all health and disease in the world is dependent on Bacteria. If we are therefore spared that customary portion of the introduction to a lecture which seeks to impress the hearer with the importance of the subject, it becomes the more necessary to give prominence from the first to the reverse side of the question; that is to say, to call special attention to the fact, that the problem presented to us can only be solved by quiet scientific examination from every possible point of view of the objects under consideration; and a study of this kind is dry rather than exciting, or to use a common expression, interesting. But this should not deter any one who is really desirous of acquiring some knowledge of our subject.

The order of our remarks will follow the natural arrangement of the subject before us; and our first task therefore will be to enquire what Bacteria are; in other words, to make ourselves acquainted with their conformation, their structure, their development, and their origin in connection with their development. Next, we have to enquire what they do, what good and what

B

harm they occasion, that is, we must study their vital processes and the effects which these produce on the objects outside of themselves.

We begin with the first question, and we will first of all bestow a moment's consideration on the name.

Bacteria, meaning rod-shaped animalcules or plantlets, from the rod-like form which many of them exhibit, are also termed Fission-fungi or Schizomycetes. The two expressions are not, strictly speaking, of the same import.

The reason of this is that the word Fungi is used in two senses. In the one it is the name for those lower flowerless plants which are devoid of chlorophyll, the green colouring matter of leaves, and hence exhibit certain definite peculiarities in the process of their nutrition. We shall speak of these peculiarities at greater length in succeeding lectures; at present we will only make the preliminary observation, that all organisms devoid of chlorophyll require already formed organic carbon-compounds for their nutrition, and cannot obtain the necessary supplies of carbon from the carbon dioxide which finds access to them. The construction of organic compounds from this substance is bound up with the presence of chlorophyll and analogous bodies.

Fungi in this sense are therefore a group characterised by definite physiological peculiarities the mark of which is the absence of chlorophyll, somewhat in the same way as birds and bats may be grouped together under the head of winged creatures.

In the other sense, that of descriptive taxonomic natural history, the term Fungi denotes a group of lower plant-forms distinguished by definite characteristics of structure and development, and recognised at once when we see a mushroom or a mould. The members of this group are all as a matter of fact devoid of chlorophyll, but they might contain chlorophyll and yet belong to this group, just as a bird may have no apparatus for flight and yet be allowed to be a bird. To these Fungi, as defined by natural history and not by physiological characters only, Bacteria are as little related in structure and development as bats are to birds; the relationship is even less, because there are a few, though only a few, true Bacteria which contain chlorophyll and decompose carbon dioxide, and which are therefore not Fungi in the physiological sense.

For these reasons we shall be more strictly correct if we speak on the present occasion of Bacteria rather than of Fission-fungi; but so long as we are quite clear as to the difference in the meaning of the two words, it is a matter of no importance which we use.

The conformation, structure, and growth of Bacteria are extremely simple, if we put out of sight certain phenomena of propagation and consider only the vegetative state.

Bacteria appear in the form of round or cylindrical rod-shaped, rarely fusiform, cells of very minute size. The diameter of the round cells or the transverse section of the cylindrical cells is in most cases about o·oo1 mm. (= 1 micromillimetre = 1 μ) or even less. The length of the cylindrical cells is 2–4 times the transverse section, rarely more. There are only a few forms with distinctly larger dimensions. Putting aside, for later consideration, the forms from the group of Beggiatoa, Crenothrix and their allies, which differ to some extent in this and other respects from the rest of the Bacteria, the greatest breadth yet observed is 4 μ, the measurement given by Van Tieghem for the rod-shaped cells of Bacillus crassus.

We are obliged to apply the term cells to those minute bodies, because they grow and divide like plant-cells, and also because all that we know of their structure agrees with the corresponding phenomena in plant-cells. It is true that their small size does not permit of our going at present very deeply into the minutiae of their structure. Cell-nuclei, for instance, have not yet been observed in them; but this is the case in many small cells of other plants of a low order of growth, especially Fungi, and till recent times it was the case with respect to all fungal cells.

Perseverance and constantly improving methods of research advance our knowledge as time goes on.

The Bacterium-cell is mainly composed of a portion of protoplasm, which in the smaller and in most also of the larger forms appears as an entirely homogeneous translucent substance, but in some of the larger forms it is also often finely granular or shows a different kind of structure, which will be further described presently. It consists, as Nencki (2) has shown in a number of cases, chiefly of peculiar albuminoid compounds (mycoprotein, anthrax-protein) which vary with the species, and its behaviour, when the usual empirical reagents are applied to it, agrees in general with that of the protoplasmic bodies of other organisms—the yellow and brownish-yellow coloration with solutions of iodine, and the absorption of, that is to say, the intense staining by, preparations of carmine and anilin dyes. Various specific differences occur in individual cases in the behaviour of the protoplasm to these colouring reagents, and supply very useful marks of distinction in certain cases which will be mentioned again on subsequent occasions.

We have already alluded to the fact that the protoplasm of certain Bacteria described by Engelmann and van Tieghem, for example, Bacillus virens, v. T., is coloured by chlorophyll, being of a uniform pale leaf-green hue. In the very large majority of cases it is colourless ; most Bacteria, not only when isolated under the microscope but also when collected into masses, have a pure or dirty-white colour, and in the latter case show various shades of tint inclining to gray or yellow, &c., which the practised observer may even apply to the determination of species. On the other hand, there are not a few Bacteria which exhibit lively colours when they are associated in masses, yellow, red, green, violet, blue, brown, &c., according to the individual. Schröter has collected together a number of such cases. How far these colours belong to the protoplasm itself or to its envelope, the cell-membrane, which will be described presently, or to both, cannot in most cases be certainly ascertained, because the individual cell is so small that it does not by itself show any indications of colour. In some comparatively large forms, those, for instance, grouped together by Zopf under the name of Beggiatoa roseo-persicina, it can be seen that the living protoplasmic body shares at least in the coloration, which in this case is a bright red. Some of the colouring matters in question have been submitted to closer examination and have even received special names, as bacterio-purpurin, &c. In their optical qualities they show various points of resemblance to anilin dyes, as is indicated by the above name; but we must not infer from this that the chemical composition is analogous.

Among other phenomena of frequent recurrence in the structure and contents of the protoplasm the starch-reaction claims special attention. Bacillus Amylobacter and Spirillum amyliferum, v. T., in certain stages of their development have this peculiarity, that a portion of their protoplasm, distinguished from the remainder by being somewhat more highly refringent, when treated with watery solution of iodine assumes an indigo-blue colour like starch-grains, or speaking more exactly like the granulose which forms a large part of their substance. The conditions under which this phenomenon makes its appearance and again disappears will be discussed at greater length below. E. Hansen's Micrococcus Pasteurianus also and usually Leptothrix buccalis show the granulose-reaction. We may also mention in this connection the occurrence of sulphur-granules in Beggiatoa, referring the reader to Lecture VIII for further particulars.

The protoplasmic body of the Bacteria is surrounded by a membrane or cell-wall. This membrane in one of the species which have been examined, Sarcina ventriculi (see Lecture XI), possesses, as far as is at present known, the qualities of typical plant-cellulose-membrane; it is firm and thin, and assumes the characteristic violet colour when treated with Schulze's solution. But in the majority of cases there is no trace of the characteristic coloration of cellulose. In single specimens scattered about in a fluid the membrane appears under the microscope as a delicate line drawn round the free surface, and forming the boundary between contiguous cells. It may even be seen distinct from the protoplasm in the larger forms by the aid of reagents which strongly contract the protoplasm and colour it at the same time without affecting the membrane, for instance alcoholic solution of iodine (see Fig. 1, p). It is plainly shown also in the formation of spores which will be described in Lecture III. This membrane, which lies close upon the protoplasm, is in certain forms at least, the species of Beggiatoa and Spirochaete for example, highly extensible and elastic, for it is seen to follow the curves often made by the elongated organism, and the protoplasm can alone be the active agent in producing these. But the membrane which thus directly covers the protoplasm is certainly in all cases only the innermost firmer layer of a gelatinous envelope surrounding the protoplasmic body. This may be seen directly in not a few forms if observed attentively under the microscope, when the cells or small aggregations of cells lie isolated in a fluid. Large masses of Bacteria are always more or less gelatinous or slimy when in a sufficiently moist condition. When the cells are dividing, the outer layers of the membrane may sometimes be seen to swell up in succession. Hence, speaking generally, we may say that the cells of Bacteria have gelatinous membranes, with a thin and comparatively firm inner layer. The consistence of the mucilage and its capability of swelling in fluids differ in different cases, changing gradually, but this point will be considered again presently at greater length.

The possession of gelatinous membranes of this kind is common to the Bacteria and to various other organisms of the lower sort, of which Nostocaceae and some Sprouting and Filamentous Fungi may be quoted as examples. In Bacteria, as in the latter plants, the gelatinous membrane has been shown in a number of forms which have been examined to consist of a carbohydrate closely related to cellulose; this is specially the case in the Bacterium of mother of vinegar and in Leuconostoc, the frog-spawn-bacterium of sugar factories. On the other hand, Nencki found that the membranes of certain putrefactive Bacteria not distinctly determined are in a great measure composed, like the proto- plasm which they enclose, of the mycoprotein mentioned on page 4. Lastly, a statement of Neisser (65) must also be men- tioned in this place ; he suspects, from the behaviour of the membrane or envelope of the Bacterium of xerosis conjunctiva in the presence of reagents, that it contains a considerable amount of fatty matter. Further investigation into these points is at all events desirable. The membranes of Cladothrix and Crenothrix which live in water are often coloured brown by the introduc- tion of compounds of iron.

Many Bacteria are capable of free movement in fluids. They rotate about their longitudinal axis, or they oscillate like a pendulum and move rapidly forwards or backwards. Search has consequently been made for organs of motion, and these are supposed to have been found in certain very slender filiform appendages, cilia or flagella, which are attached singly or in pairs to the extremities of rod-shaped Bacteria. Such cilia are present in many relatively large cells not belonging to the Bacteria, and endowed like them with the power of free movement in fluids, the swarm-cells, for example, and swarm-spores of many Algae and some Fungi. In these cases the cilia oscillate rapidly as long as the movement continues, causing rotation round the longitudinal axis, and may consequently be considered to be the active organs of motion. In the swarm-cells of the Algae they are processes, projections as it were, from the surface of the protoplasmic body, and belong therefore to the protoplasm. When the protoplasm is surrounded by a membrane, the cilia pass out through openings in the membrane. But no such characteristic structural conditions have been observed in the Bacteria. Delicate thread-like processes have certainly been observed occasionally at the points above-mentioned in coloured specimens which have been exposed to desiccation. That they are really there and not, or at least not always, in the imagination of the observer only, is proved by the fact that they appear in photographs. But in an overwhelming majority of cases no cilia can be seen, though the Bacteria are capable of independent movement and are examined with the best optical aids after being killed and coloured. Where they are found, they are as van Tieghem rightly says, not processes of the protoplasmic body, but belong to the membrane, as is shown by their behaviour with reagents, and must therefore be considered to be thread-like extensions of the soft gelatinous membrane-layers. They have accordingly nothing in common with the cilia of swarm-spores of the Algae, and cannot therefore be regarded as organs of motion, since it was only from the analogy of the cilia in the Algae that this function was inferred. Such is the state of the case at least in the great majority of species. Whether there are any exceptional cases must be determined by further investigation. It should be added, that among lower organisms there are some comparatively large forms, the Oscillatorieae, for example, the near relatives according to our present knowledge of the Bacteria—a point to be further considered below—which show similar movements, though no cilia or other distinct organs of motion have been observed in them. It follows that analogy does not require the discovery of cilia in the Bacteria.

Vegetating Bacterium-cells multiply by successive division, each cell forming two daughter-cells. When a cell has reached a certain size, a fine transverse line makes its appearance in it, dividing the cell into two equal parts. This line is subsequently shown by its gelatinous swelling to be the commencement of a cell-membrane. This agrees with the phenomena observed in the divisions of larger plant-cells, and there is nothing to prevent our assuming that the details of the process of division, which the minuteness of the object makes it impossible to observe directly, are the same in both cases.

It must be acknowledged that the transverse wall which appears as the cell divides is often so delicate as easily to escape observation, and becomes visible only under the influence of reagents which give a deep colour to the protoplasm and make it shrink, especially alcoholic solution of iodine. This must not be forgotten in determining the length of cells.

The successive bipartitions are either all in the same direction, and the transverse walls are therefore parallel; or more rarely the walls lie in two or three directions in space, so that they successively cut one another, and may actually cross at a right angle.

II.

Cell-forms, cell-unions, and cell-groupings.

SINGLE Bacterium-cells, the simple structure of which has been considered in the preceding chapter, may appear in very various forms, the variety depending partly on their own shape and on that of their simplest aggregations, partly on whether they are united into larger aggregations or not, and on the peculiar characters of these aggregations.

1. The shape of the individual cells and their simplest genetic combinations give rise to the distinction into round-celled, and straight or spirally twisted rod-like forms. A billiard-ball, a lead pencil, and a cork-screw, so exactly illustrate these three chief forms, that no one requires for his instruction in this case the costly models which are offered for sale. The figures on subsequent pages, which will be examined more closely in later lectures, will for the present give a sufficiently clear idea of the matter.

These forms have received a variety of names in the course of the development of our knowledge. The round forms are at present most commonly known as Cocci (Figs. 3, 4), and are spoken of as Micrococci or Macrococci according to their size, or as Diplococci when they still remain united in pairs after a bipartition; earlier writers called them monads, a name which they applied to a variety of heterogeneous objects.

The straight rod-forms (Figs. 1, 2) have received the special name of rods, Bacteria, from the earlier writers. Short or long rods and other terms are obvious designations for subordinate peculiarities of shape, but have no other value.

The screw or cork-screw forms are termed Spirilla, Spirochaetae. Those which are only slightly curved, that is, which form a portion only of a turn of the screw, being intermediate between the two preceding categories, have been called by Cohn Vibriones in accordance with the nomenclature of older authors. It is well that we should understand clearly that these and other names, which will be mentioned presently, are only used to define the shapes of the organisms. It would indeed be better to give them proper names expressive of their outward appearance, and to use terms like sphere, screw; and it is to be hoped too that the jargon which prevails at present, especially in medical literature, will gradually be replaced by a rational terminology.

The cocci and rod-forms are sometimes liable to a peculiar deviation from their ordinary shape; single cells, lying between other cells which remain true to one of the typical forms described above, swell into broadly fusiform or spherical or oval vesicles several times larger than the typical cells. This has been observed in species of Bacillus, Cladothrix, &c., and with special frequency in the Micrococcus of mother of vinegar. There is some ground for assuming, though further proof is required, that these swollen forms are the products of diseased development, indications of retrogression and involution, and they were therefore termed by Nägeli and Buchner involution-forms (see Fig. 10).

2. According to the nature of the union or want of union of the cells, we must first of all distinguish between the forms in which genetic union and arrangement is maintained after successive bipartitions, and those in which it is severed or displaced.

When the cells continue united together in the connected sequence of the divisions we have—

a. The cells arranged in rows in the direction of the successive divisions.   From their thread-like form these cell-rows
(Fig. 2, &c.) are termed filaments in accordance with the
traditional terminology; a strange confusion of ideas led to
their being also called pseudo-filaments, objects which look like,
but are not real, filaments.

It is obvious after the foregoing remarks, first, that these
filaments must be of different shape according as the individual
cells are round or of some other form; secondly, that the
length of the filaments, measured by the number of cells, may
be very various.   It may be said specially of the rod-like and
spiral forms, that the cells usually remain united into short
rows in such a manner that the rod or spiral is actually
composed of more than one cell, and then after a definite
increase of the entire length and of the number of the segment-
cells is divided into two at the oldest points of division.   The
words Leptothrix, Mycothrix, and others designate the longer
filamentous forms.

b. Cells united together and arranged genetically in a simple
surface, or as a body of three dimensions, are of less frequent
occurrence, as has been already said; Zopf's Bacterium meris-
mopoedioides may be given as an example of the former kind,
of the latter the cube-shaped cell-packets of Sarcina ventriculi
(see Fig. 14).

By the side of these phenomena of genetic union and variously
combined with them appears a series of groupings, as they may
be briefly termed, which owe their character to a great extent to
the mass, cohesion, and other specific qualities of the gelatinous
membranes as these are formed, and next and in combination
with these, to their own very various specific peculiarities, which
cannot, as a rule, be shortly defined; some explanation of the
latter, though unfortunately only an imperfect one, will be given
further on when we are considering vital processes.   The nature
also, and especially the state of aggregation of the substratum,
may in certain circumstances have an influence on the grouping.

Thinness of the gelatinous membranes and a high degree of capacity for swelling reaching to deliquescence will cause the separation of cells or of the simplest cell-unions from one another when growing in a fluid.   Thick cell-membranes, and a narrowly limited capacity for swelling in the gelatinous sub-stance, will keep the cells united together in compact gelatinous masses in the same fluid.   These, which are the extreme con-ditions, are actually found in nature, and all kinds of intermediate states also occur.   The firmer gelatinous masses (see Fig. 3) are called by the old name Palmella, or by the more recent name now commonly used, Zoogloea.   The less sharply defined Zoogloeae, as they may be shortly described, may naturally be termed swarms.   It depends on their specific gravity whether a Zoogloea or a swarm will float on the surface or sink to the bottom of the same fluid, while their general outline and the grouping of the separate aggregates which compose them will be fashioned in accordance with their further specific qualities.

To illustrate this point in passing by a few examples,— let us take three flasks containing a similar 8–10 per cent. solution of grape-sugar and extract of meat in water.   In one flask the fluid is pretty uniformly clouded with the short motile rods of Bacillus Amylobacter.   In the second the surface of the slightly clouded fluid is covered with a thick wrinkled scum, dry on the upper surface, which is formed by Bacillus subtilis, the so-called hay-bacillus.   In the third the filaments of Bacillus Anthracis, the Bacillus of anthrax which in other respects re-sembles B. subtilis, form a flocculent deposit at the bottom of the clear fluid.   We can scarcely call this deposit by the name of Zoogloea, we may perhaps call it a swarm.   The hay-bacillus-scum is properly a Zoogloea with a special characteristic form.   Formations more or less like it are found often enough in fluids containing decomposable organic bodies.   Highly characteristic Zoogloeae developed in a fluid are the frog-spawn-Bacterium of sugar-factories and the Bacterium of kefir.   The former is a round-celled organism, Leuconostoc, with a thick compact gelatinous envelope which may fill entire vats with a substance looking like frogs' spawn, and which will be considered again in a later lecture; the term kefir-grains is applied to the bodies employed by the inhabitants of the Caucasus in the preparation from milk of a sourish beverage rich in carbonic acid. The kefir-grains are in the fresh living state white bodies, usually of irregularly roundish form, equal to or exceeding a walnut in size. They have their surface crisped with blunt projections. and furrowed like a cauliflower; they are of a firm toughly gelatinous consistence, becoming cartilaginous, brittle, and of a yellow colour when dried, and are chiefly composed of a rod-shaped Bacterium. The small rods are for the most part united together into filaments, which are closely interwoven in countless zig-zags and firmly connected together by their tough gelatinous membranes. It must also be observed, that the Bacterium-filaments are not the only constituents of the granules; numerous groups of a Sprouting Fungus like the yeast-fungus of beer are enclosed between them, especially in the periphery, living and growing in common with the Bacterium, but in much smaller quantity, and taking only a passive part in the formation of Zoogloeae.

If Bacteria grow not in fluids but on some solid substance which is only wet or moist, the grouping into Zoogloeae is a frequent phenomenon even in those forms which separate from one another in larger amounts of fluid owing to the deliquescence of their gelatinous envelopes. The more limited supply of water on the merely damp substratum is not sufficient to make this gelatinous substance swell up to the point of deliquescence. On decaying potatoes, turnips, and similar substances we may often see small lumps of gelatinous matter of a white or yellowish tint or of some other shade of colour, and composed of these aggregations of Bacteria. These lumps deliquesce in water. We have a special instance of the kind in the often-described occurrence of the blood-portent, Micrococcus prodigiosus (Monas prodigiosa of Ehrenberg), producing on substances rich in starch, such as dressed potatoes, bread, rice, or wafers, moist blood-red spots which sometimes spread rapidly and widely. Their colour has given rise to a variety of superstitious notions when they have appeared unexpectedly on objects of household use. They consist of one of the chromogenous Bacteria which have been already mentioned.

It was stated above that the grouping of different forms is different in the same fluid, and in like manner the conformation of the Zoogloeae on solid substances shows manifold variation of forms which differ in other respects also.

These various facts connected with the grouping are themselves calculated to afford very valuable marks for characterising and distinguishing forms, the more valuable indeed in proportion to the difficulty oftentimes of discriminating individual cells of such minute size under the microscope. It is precisely in the phenomena of grouping that specific peculiarities of conformation best display themselves, being collected together, as it were, in larger quantity; these characters must indeed be present in the single cell, but, with the means at present at our disposal, it is difficult or even impossible to recognise them there. But there is nothing peculiar in this. There are many cells of gigantic size in comparison with the Bacteria and highly differentiated, of which we cannot say with certainty when we see them by themselves whether they belong to a lily or a tulip. But in their natural connection or grouping some of them form a lily, others a tulip, and by this we know that they are different.

III.

Course of development. Endosporous and Arthrosporous Bacteria.

THE different conformations and groupings described in the preceding lectures indicate primarily nothing more than definite forms of one phenomenon marked in each case by a distinct name, such as present themselves at any moment of observation, and without reference to their origin or future destination. They are forms of the vegetative development, growth-forms as they may be shortly termed, and correspond to those which in the higher plants are designated by the words tree, shrub, bulbous plant, and the like. Forms which are determined only by their conformation correspond indeed only to separate members of a particular growth, such as woody stem, tendril, tuber, bulb, &c.

If we wish to know the significance of a tendril or a bulb in the chain of phenomena, or indeed that of any other form of living creature, we must answer the above questions of its origin and destination, or, to use the customary form of words, we must learn the course of its development. For every form of living being taken at any one moment of time, though it may be present in millions of specimens, is only a member of a chain of periodic movements which coincide with a regular alternation of forms. If therefore we wish for a more intimate acquaintance with Bacteria, we must proceed to enquire into their course of development.

As far as our present knowledge goes, this development is not quite the same in all cases. We must distinguish two groups, one of which contains the Endosporous, the other the Arthrosporous Bacteria.

The former group consists of a number of straight rod-forms which will here receive the special name of Bacillus, and a few screw-twisted Spirilla. The phenomena, so far as they are known, are essentially the same in both, and they have now to be described in detail in the case of Bacillus. See Fig. 1.

The Bacilli in the highest state of vegetative development are rod-shaped or shortly cylindrical cells with the characters already described, which either remain isolated or are united together into unicellular rods or longer filaments; they are motile or motionless, and display active growth and division (Fig. 1, *a–c*). Both growth and division at length come to an end, and then begins the formation of peculiar organs of reproduction—spores. This process begins at the point to which it has been followed furthest back with the appearance of a comparatively very minute point-like granule in the protoplasm of a hitherto vegetative cell. This granule increases in volume and soon presents the appearance of an elongated or round, highly refringent, sharply-defined body, which attains its ultimate size rapidly, sometimes in a few hours, and is then the mature spore (Fig. 1, *d–f*). The spore always remains smaller than its mother-cell, the protoplasm and other contents of which disappear with the growth of the spore, being doubtless consumed for its benefit, until at length the spore is seen suspended in a pellucid substance inside the delicate membrane of the mother-cell (Fig. 1, *r*, h_1).

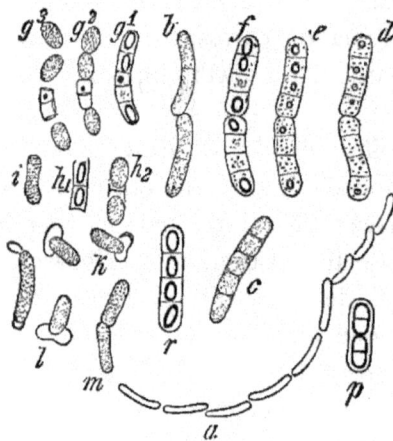

Fig. 1.

Fig. 1. Bacillus Megaterium. *a* outline sketch of a chain of rods in active vegetation and motion. *b* pair of rods in active vegetation and motion. *p* a 4-celled rod in this stage after treatment with alcoholic solution of iodine. *c* 5-celled rod in the first stage of preparation for forming spores. *d–f* successive states of a spore-forming pair of rods, *d* about 2 o'clock

The details of these processes disclose sundry variations of diagnostic value, especially in connection with the shape. In Bacillus Megaterium, B. Anthracis, B. subtilis, for example, the sporogenous cell does not differ in shape from the vegetative cell, but in the two latter the mature spore is much shorter; in B. Anthracis it is slightly narrower, in B. subtilis often rather broader than the mother-cell, in B. Megaterium it is a little shorter but much narrower than the comparatively short mother-cell (cf. Figs. 1 and 2).

In other species the spores are much smaller in every direction than the mother-cell, and the latter is distinguished from the cylindrical vegetative cell before or during the formation of the spore by swelling into a permanent fusiform or ovoid shape, either over the entire area of the cell or at the spot where the spore lies, and which is then usually at one extremity of the cell. In the latter case, and also when cells that are still cylindrical are attached on one side to a mother-cell which has swollen up all over, the forms are produced which were once known as capitate Bacteria, cylindrical Bacteria with a capitate sporogenous swelling at the extremity. Examples of this kind are Bacillus Amylobacter (Fig. 13), B. Ulna, and some others.

In Bacillus Amylobacter and Spirillum amyliferum, v. Tiegh., the appearance of the spore is preceded by the formation of granulose described above, and the spot where the spore

p.m., *e* about one hour later, *f* one hour later than *e*. The spores in *f* were mature by evening; no others were formed; the one apparently commenced in the third upper cell of *d* and *e* disappeared; the cells in *f* which did not contain spores were dead by 9 p.m. *r* a four-celled rod with ripe spores. *g*[1] five-celled rod with three ripe spores placed in a nutrient solution after drying for several days, at 12.30 p.m.; *g*[2] the same specimen about 1.30 p.m.; *g*[3] about 4 p.m. *h*₁ two dried spores with the membrane of the mother-cell placed in a nutrient solution, about 11.45 a.m.; *h*₂ the same specimen about 12.30 p.m. *i, k, l* later stages of germination explained in the text on p. 21. *m* rod dividing transversely, grown from a spore placed eight hours before in a nutrient solution. *a* magnified 250 times; the other figures 600 times.

begins to be formed is marked by the absence of granulos
This spot looks in solution of iodine like a notch of a pa
yellowish colour, occupying one extremity of the rod whic
elsewhere tends to be blue, and is moreover distinguished l
its lower power of refraction even before the use of a reager
As the spore grows in size the granulose disappears. Accor
ing to Prazmowski the granulose is not always present befo
the formation of the spore, even in Bacillus Amylobacter. 1
other Bacilli, the three, for example, just previously named,
has never been observed; their protoplasm either remains u
changed before spore-formation, or at most becomes a litt
less transparent and in larger forms more evidently fine
granular.

A mother-cell, so far as can be positively stated, never pr
duces more than a single spore. This can be determined wi
certainty in almost all cases, and the few accounts which ha
been given of the formation of two spores in a single cell a
doubtful, being unaccompanied by any guarantee that tl
boundaries of adjoining cells have not been overlooked
errors of other kinds admitted. I must, however, add that ε
exception to the prevailing rule has recently come under n
own observation in the case of a species nearly allied to Bacillι
Amylobacter (see Lecture IX), which usually follows the rul
but does also sometimes contain two spores in a cell which h
swollen and become broadly fusiform. I have not yet succeedε
in observing the further development of the twin spores.

In cultures formation of spores usually takes place whε
other growth comes to an end because the substratum is ι
longer adapted to maintain it, being either exhausted, as ν
are in the habit of saying, or impregnated with the produc
of decomposition which are unfavourable to vegetative develoι
ment. Formation of spores then spreads rapidly through tl
larger number of the cells and through the special aggregatior
if the particular form is present in abundance. Some of the
it is true do not produce spores, in some the process begiι

but is not completed. All cells which do not take part in the normal formation of spores ultimately die and are decomposed, if they are not transferred in good time to a fresh substratum.

In other Bacilli, as B. Amylobacter, the procedure is different. Here spore-formation begins in single cells and spreads by degrees to more and more of them, while a number of other cells continue to vegetate and divide. We cannot therefore regard the unsuitableness of the substratum to the vegetative process as the cause which generally determines the formation of spores.

By spores are usually meant such cells as are delimited from a plant to develope again under favourable conditions into a new vegetating plant. The commencement of this latter process is termed germination. The bodies to which we have here given the name of spores are so called because their behaviour corresponds to that of germinating spores. As soon as they are fully grown, that is, as soon as they are ripe, the membrane of the mother-cell dissolves gradually or swells, and the spores are thus set at liberty, retaining the characters which have been already described; that is, they are round, ovoid, or rod-shaped, according to the species, rarely of some other shape, with a dark outline and usually colourless, but with a peculiar bluish glistening appearance; according to Cohn the spore of Bacillus erythrosporus is tinged with red. Round the dark outline may often be perceived a very pale and evidently soft gelatinous envelope, which either covers the spore uniformly all round, or is thicker at the two extremities and drawn out into processes.

Germination shows that the spore is a cell surrounded by a thin but very firm membrane, defined by the dark outline inside the gelatinous envelope. Germination begins when the ripe spore is subjected to the conditions favourable to the vegetation of the species, supply of water, suitable nutriment, and favourable temperature. As it begins the spore loses its high refringent power, its lustre and dark outline; it assumes the appearance of a vegetative cell, and grows at the same time to the size and shape of the vegetative cell from which it sprang. With the completion of this process movement begins in the motile species, and this is followed by the growth, division, and grouping which have been described above as occurring in the vegetative stages, and which at length come to an end with a fresh formation of spores. In many cases a few hours only intervene between the first observable commencement of germination and active vegetative growth. See above, Fig. 1, *h–m*.

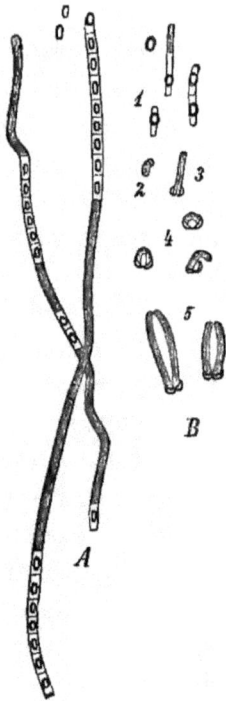

With the first increase in size a membrane is often seen to split and rise from off the surface of the growing cell, being evidently lifted from off it by a swelling gelatinous outer layer surrounding the new membrane of the cell. The rent through the membrane is in the direction of the length, or across the middle, according to the species. The former is the case according to Prazmowski in Bacillus Amylobacter, and it occurs also in other species. The latter has been observed in

Fig. 2.

Fig. 2. *A* Bacillus Anthracis. Two filaments partly in an advanced stage of spore-formation; above them two ripe spores escaped from the cells. From a culture on a microscope-slide in a solution of meat-extract. The spores are drawn a little too narrow; they are nearly as broad as the breadth of the mother-cell. *B* Bacillus subtilis. 1 fragments of filaments with ripe spores. 2 commencement of germination of spore; the outer wall torn transversely. 3 young rod projecting from the spore in the usual transverse position. 4 germ-rods bent into the shape of a horse-shoe, one afterwards with one extremity released. 5 germ-rods already grown to a considerable size but with both extremities still fixed in the spore-membrane. All magnified 600 times.

B. Megaterium (Fig. 1) and B. subtilis (Fig. 2, *B*); the transverse rent either extends quite across, so that half the membrane is placed like a cap on each extremity of the cell, or the halves remain attached on one side, so that the growing cell must emerge from a gaping cleft (Fig. 1, *h–l*). The ruptured membrane is usually delicate and pale. In B. subtilis only it retains at first the lustre and dark outline of the spore before germination, and hence it is probable that these phenomena have their origin in the membrane. Sooner or later the membrane thus torn from the cell swells and disappears. It may be owing to the very early period at which the swelling sets in that sometimes, as, for example, in B. Megaterium and B. Amylobacter, the removal of the membrane is not perceptible in one germinating spore, while it is clearly seen in others, and that in other species, as in B. Anthracis, no removal of the membrane takes place at all.

The longitudinal growth of the first cell in germination has always the same direction in space as the longitudinal axis of the spore or spore-mother-cell. This is the case also in Bacillus subtilis, which appears at first sight to behave differently in this respect. In this species the first rod-shaped germ-cell usually emerges from the open transverse rent in the spore-membrane in such a manner that its longitudinal axis crosses that of the spore at a right angle, but this is not caused by a corresponding divergence of the longitudinal growth, but by the circumstance that when the germ-cell has attained a certain length it bends through about 90°, and thus projects on one side at a right angle from the rent in the membrane. The bending of the germ-cell is evidently caused by the resistance offered to the elongation of the cell by the spore-membrane, which in this species is highly elastic and is always ruptured on one side. When growth is very rapid the two extremities of the young rod may remain fixed in the membrane, and in that case the middle portion projects in a curve from the aperture. It is not till a later period, when the rods have begun to divide and separate into daughter-rods, that the latter straighten themselves out.

Endogenous spore-formation, as it has now been described, that is, formation of spores taking place inside the previously vegetative cell, sharply distinguishes the endosporous forms from the rest of the Bacteria, which we have termed arthrosporous. The name is intended to indicate the fact, that in these forms members of an aggregation or of a series of united generations of vegetative cells separating from the rest assume the character of spores immediately without previous endogenous rejuvenescence, that is, they may become the origins of new vegetative generations. In a number of the forms comprised in this division, a more or less distinct morphological difference may be observed between vegetative cells and spores; in others, as far as we know at present, no such distinction is to be found.

Simple examples of the former kind are supplied by Leuconostoc mentioned above and by Bacterium Zopfii, Kurth. The former (Fig. 3) consists, according to van Tieghem's description, of curved bead-like rows of small round cells with firm gelatinous coats united together in large numbers into Zoogloeae (Fig. 3, *a*, *b*). A large portion of the cells dies at the end of the vegetative period when the nutrient substratum is exhausted. On the other hand single cells irregularly distributed in the rows become somewhat larger than the rest, acquire a more distinct outline, that is, become thicker-walled, and their protoplasm grows darker (Fig. 3, *c*). They at length become free by the deliquescence of the gelatinous envelopes, and may claim the name of spores, because when placed in a fresh nutrient solution they develope into new rows of beads like those of the mother-plant (Fig. 3, *d–h*).

Bacterium Zopfii was originally found by Kurth in the intestinal canal of fowls, and then cultivated partly in gelatine partly in suitable watery solutions. In the fresh substratum the Bacterium vegetates at first in the rod-form. In the gelatine the rods continue united together into large filaments often twisted

into a coil; in the fluid short and motionless filaments are
formed only at a temperature of more than 35° C.; at 20° C. the
filaments separate into motile rods. At the close of their vegetation when the substratum is exhausted, the rods fall asunder

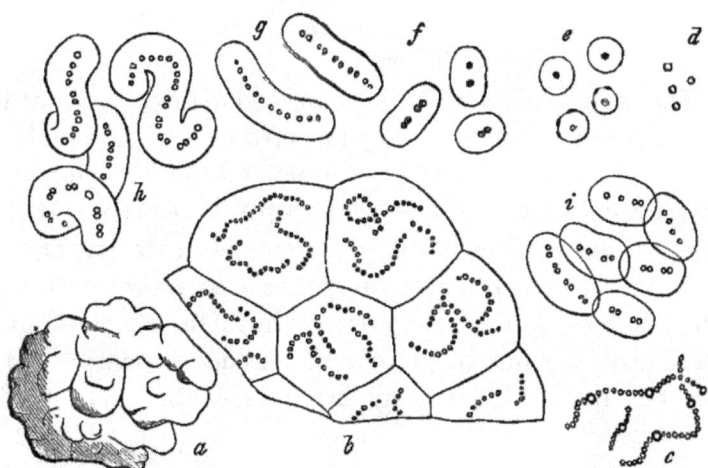

Fig. 3.

into short roundish cells, and these again may be termed spores
since in a fresh substratum they develope into new rods or
filaments.

Though their course of development is more complex yet the
phenomena observed in Crenothrix, Cladothrix, and Beggiatoa,
if Zopf's description is correct, closely resemble those just described. They will be noticed again below in Lecture VIII.

Fig. 3. Leuconostoc mesenterioides, Cienkowski. *a* sketch of a Zoogloea.
b section through a full-grown Zoogloea just before the commencement of
spore-formation. *c* filaments with spores from an older specimen.
d isolated ripe spores. *e–i* successive products of germination of spores
sown in a nutrient solution; sequence of development according to the
letters. In *e* the two lower specimens show the fragments of the ruptured
spore-membrane on the outer surface of the gelatinous envelope indicated by
dark strokes. *i* portion of a gelatinous body from *h* broken up into short
members, which have been separated from one another by pressure. After van
Tieghem (Ann. d. sc. nat. sér. 6, vii). *a* natural size, *b–i* magn. 520 times.

Examples of the other and simpler kind of arthrosporous forms are to be found, according to our present knowledge, in the forms described under the name of Micrococcus (Fig. 4). Each vegetative cell may at any moment begin to form a new series of vegetative cells ; there is no distinction between specifically reproductive and vegetative cells.

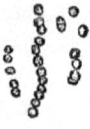

Fig. 4.

The entire distinction between endosporous and arthrosporous Bacteria is required by the present state of our knowledge. It remains to be seen whether and to what extent it will be maintained. Our knowledge is at present still so incomplete that we must on the one hand regard the discovery of endogenous formation of spores as possible or probable in forms where they are hitherto unknown, even in the simplest Micrococci, and I say this to prevent all misunderstanding; and on the other hand we cannot say that facts will not come to light in course of time which will do away with any sharp separation between the two divisions.

IV.

Species of Bacteria. Distinct species denied. The grounds for this denial insufficient. Method of investigation. Relationships of the Bacteria and their position in the system.

HAVING now made ourselves acquainted with the course of development in Bacteria in its main features, we proceed to consider the much-debated question whether there are specifically distinct forms, species of Bacteria, as these terms are used in descriptive natural history, and how many such species can be determined. Species are determined by the course of development. By the term species we mean the sum

Fig. 4. Micrococcus Ureae, Cohn, from putrifying urine. Single cells and cells united in rows (= Streptococcus). Magn. 1100 times.

total of the separate individuals and generations which, during
the time afforded for observation, exhibit the same periodically
repeated course of development within certain empirically determined limits of variation.   We judge of the course of development by the forms which make their appearance in it one after
another.   These are the marks by which we recognise and distinguish species.   In the higher plants and animals we are
in the habit of taking the marks chiefly from a single section of
the development, namely, from the one in which they are most
distinctly shown.   We distinguish birds better by their feathers
than, for instance, by their eggs.   This abbreviated method of
distinguishing is convenient, wherever one section of the development is so pregnant as to make the consideration of the
rest unnecessary.   But this is not always the case.   The simpler
the forms of an organism are, the larger must be the portions
of development requisite for characterising and distinguishing it,
and the demand is still greater when we have to compare the
entire course of the development of the species, to use the same
figure, from the ovum of the first to the ovum of the next generation.   We are pleased if we succeed in this way in finding any
single mark to serve our purpose, but we must not be too con-
fident of finding one.

Experience has taught us that different species may behave
very differently in respect of the forms which make their appearance successively in their course of development.   In some the
same forms constantly recur one after another with comparatively
small individual variations.   These may be termed monomorphic
species.   Most of the common higher plants and animals are
examples of this, and also many of the lower and simpler kinds.
They can be readily distinguished after a little experience even
by single portions from the general development.   We can recognise a horse-chestnut, for example, by each individual leaf
plucked from the tree.

Other species are pleomorphic and may appear in very
unlike forms even in the same segments of the development,

partly from the effect of external causes which are known and
may be varied at pleasure in our experiments, partly from
internal causes which cannot at present be analysed. The
white mulberry-tree, for example, in contrast to the horse-
chestnut just mentioned produces foliage-leaves very unlike each
other and with no certain rule of succession, some simply cordate, others deeply notched and lobed. We should not recognise the species by a leaf of the latter kind, if we had before
only happened to have seen the cordate leaves. This occurs
frequently and to a still greater extent in the lower plants,
though they need by no means belong, like the Bacteria, to the
simplest and smallest forms. Many of the larger Fungi, for example, the forms of Mucor, and green Algae, such as Hydrodictyon
and the remarkably pleomorphous Botrydium granulatum, exhibit
phenomena of this kind in a very striking manner, especially
when it further happens, as it often does happen in similar
plants, that the successive members of the development do
not continue in prolonged connection with each other, like the
leaves of the mulberry, but separate and vegetate apart from
one another. In this case if we happen to find the objects
separate and alone, and are accustomed from our experience
of the chestnut always to judge of a species by the individual
form, we fall into mistakes such as the history of botanical study
can supply in great abundance. But if we observe how each
form developes and how it originated, we perceive that they
have all the same course, the same origin, and the same return
to similar beginnings, or as we may say, conclusions of the
development.

The pleomorphous species therefore differ from the relatively
monomorphous species only in the greater number of forms
and in the greater amount of differentiation in the course of
development; the qualities of the species are apportioned in
equal measure in the one as in the other.

As regards then the species of Bacteria two views have been
promulgated, which in their extreme form differ much from one

another. According to the one view their case is the same as
that of all organisms other than Bacteria, that is, of all other
plants and animals; like these they are distinguished into
species. This was accepted as a matter of course by the earlier
observers from the first discovery of the Bacteria by Leeuwenhoek (3), to the more careful and extended observations of these
organisms which was undertaken by Ferdinand Cohn (4) at the
beginning of the period from 1860 to 1870. Cohn, following
in the steps of his predecessors, especially Ehrenberg (5), endeavoured to give a general view and classification of the forms
which had become known to himself and others. It was important to arrange the material in hand and waiting further ela-
boration in some provisional manner, and to do this it was
either allowable or necessary to start from the assumption,
which certainly required to be proved, that a species was always
characterised by a definite form, as is the case with the above-
mentioned comparatively monomorphous kinds. The species
were therefore distinguished by their shape and growth-form, with
some help from their effects on the substratum, and then further
classified. The names Coccus, Spirillum, Spirochaete, &c., applied
above to growth-forms corresponding to such terms as tree and
shrub, were used as names for definite natural genera like birch,
chestnut, &c.; such genera we may accordingly therefore term
form-genera. Whether these form-genera and form-species did
or did not really coincide in all points with natural genera and
real natural species, was expressly left undecided by Cohn and
reserved for further investigation.

Cohn's view as expressed in his provisional classification was
opposed by other writers, who went so far as to deny that there
were any species of Bacteria. They considered that the observed forms proceeded alternately from one another, the one
being convertible into the other with a change in the conditions
of life, and that this change might be accompanied by a corresponding change in the effects on the substratum, though this
point does not, strictly speaking, belong to the subject which we

are considering. Full expression was given to this view by Billroth (6) in 1874 in a lengthy publication, in which he includes all the forms which he had examined, and they were many and various, in one species which he names Coccobacteria septica. Nägeli (7) and his school have supported the same views since 1877. Nägeli indeed expresses his opinion on the one hand with circumspection and reserve, saying that he finds no necessity for separating the thousands of Bacterium-forms which have come under his observation even into two species, but that it would be rash to speak decidedly on a subject that is so imperfectly explored. On the other hand he goes so far as to say: If my view is correct, the same species in the course of generations assumes a variety of morphologically and physiologically dissimilar forms one after another, which in the course of years and decades of years at one time turn milk sour, at another give rise to butyric acid in sauerkraut, or to ropiness in wine, or to putrefaction in albumen, or decompose urine, or impart a red stain to food-material containing starch, or produce typhus, relapsing fever, cholera, or malarial fever.

In presence of this statement of opinion our practical interests require that we should obtain a decided answer to the question of species which we are here considering, for it certainly is not a matter of indifference in medical practice, for example, whether a Bacterium which is everywhere present in sour milk or in other objects of food, but without being injurious to health, is capable or not of being changed at any moment into a form which produces typhus or cholera. The scientific interest certainly demands that the question should be set at rest.

It may safely be maintained that continued investigation has at length arrived at a decision and it is this, that there is no difference as regards the existence of species and their determination between this and any other portion of the domain of Natural History.

Species may be distinguished provided we follow the course of development with sufficient attention. Some which are familiar to us through the researches of Brefeld, van Tieghem, Koch, and Prazmowski are comparatively monomorphous; they make their appearance in the vegetative segments of their development as a rule in the same forms as regards their shape, growth and grouping. Others show a greater amount of variation in these respects; they display the phenomena of pleomorphy in varying degrees. Among the endosporous Bacilli described above Bacillus Megaterium is a particularly good example of monomorphy. A motile rod is developed from the spore and gives rise as it grows to successive similar generations of rods, until these at length proceed to the formation of fresh spores (Fig. 1).

Bacillus subtilis when growing normally in a fluid differs to some extent from B. Megaterium; successive generations of rods moving about in the fluid proceed from the germinating spores, but the later generations which proceed from them remain united into long filaments and are without motion, being grouped together and forming on the surface the Zoogloea-membrane mentioned on page 12. In this state they then form fresh spores. Here therefore we have a small amount of pleomorphy, two, or reckoning the spores, three distinct forms, and in a sequence also which is regularly repeated from one spore-generation to another. Moreover the special conditions of shape and size always remain the same within certain limits of variation, for variations in the direction indicated certainly occur in this case as they do everywhere in the organic world. Stunted forms may also be met with. I have for instance repeatedly observed some of the rods in a group of Bacillus Megaterium, in circumstances unfavourable to its nutrition, separate into its cells which were themselves already short, and these cells round themselves off and in this way represent what may be termed Cocci. Other unusual forms also made their appearance with the Cocci. There was scarcely any or no formation of spores. Under improved food-conditions these stunted forms reverted to the normal state.

The arthrosporous species, Crenothrix and Beggiatoa, which have been mentioned above, are particularly striking examples of pleomorphous species, if the accounts of them which we possess are correct (see Lecture VIII). But it is in these very forms that the course of development, as we have already pointed out, has not been so completely followed out and so clearly explained as to exclude the possibility, that the apparently irregular pleomorphism is due in these cases to the admixture sometimes of several less pleomorphous species. And even if we accept very extreme statements with regard to Bacteria, it is nevertheless true that the most pleomorphous Bacteria show a very high degree of uniformity when compared with the lower plants mentioned on page 26, such as Botrydium, Hydrodictyon, and many others.

Any one not familiar with the subjects and investigations in question will be inclined to ask how there can be such a profound difference of opinion as that between the negation and affirmation of the existence of species in Bacteria. The answer is, that the difference has its origin in the differences and to some extent in the mistakes in the method of investigation. I do not use the word method here in the customary sense of manual skill and practical contrivances in investigation, but to express the course of procedure in examining and judging of the observed phenomena.

Species, as is acknowledged and as has been already pointed out above, can only be determined and recognised in and through the course of development, and this consists in the successive development of forms, one from another. The forms which appear later in the series proceed from the earlier forms, as parts of them, and are therefore at every moment in unbroken continuity with them, even when subsequently separated from them. The proof that they all belong to one and the same course of development can only therefore be established by proving this continuity. The attempt to establish it in any other way, for example, by ever so careful observation of the

forms which make their appearance one after another at the
same spot, or by the construction of a hypothetical series of de-
velopments by the most exact and ingenious comparison of
these forms, involves a logical fallacy. We distinguish, for instance, a species of wheat by its seed, its stem and leaves, its
flowers and fruits, and we know that these proceed alternately
from one another ; but the latter fact we know only by observing that the one of these members arises as part of one of
the others, and by observing also how this happens. This is
the only reason why we consider the grain of wheat to belong
to the wheat-plant, whether it is attached to it or has fallen to
the ground or lies thrashed out on the floor of the granary.
That the stem with the leaves belongs to the grain we know by
observing its origin as part of the grain, not because we have
seen wheat-plants growing where wheat was sown ; weeds may
grow at the same spot along with the wheat.

This mode of viewing the matter sounds trivial ; its truth will
seem obvious to every one, and rightly so ; and yet it cannot be
too often repeated, for the logic which it is intended to illustrate
is being constantly disregarded, and a mass of confusion has
been the result of this neglect. This may be shown by means
of the very example which we have chosen, for less than fifty
years ago it was maintained that all sorts of weeds were produced from the seed of the wheat-plant, and people (8) in other
respects well-educated and intelligent believed that this was
possible, because these weeds sprang up in the spots where
wheat had been sown. But whoever examines at the right place
finds that either wheat or nothing grows from the wheat-grain,
that the weed springs only from the seed of the species of
weed which may happen to be present, and that where
the weed grows up instead of or with the wheat, its seed has
found its way by some means to the place where the wheat
was sown.

Notions and mistakes like these in the case of the wheat-
plant have appeared again and again in connection with smaller

organisms, such as Algae and Fungi, both those of the larger kinds and the microscopically minute. The separate species were imperfectly known, and different ones were brought into genetic connection with one another, because observation of continuity was omitted or imperfectly made, and in its place was substituted the observation of the succession in time of forms at the same spot, or the comparison of them as they made their appearance there together.

The smaller and simpler the forms, the greater certainly is the difficulty of satisfying our logical demand, and the greater the attention which must be paid to it. In small forms consisting of isolated cells of no very marked shape, such as some of the lower Fungi and the Bacteria, we must observe carefully whether the sowing contains the germs of a single species or of several mixed together. The latter is very frequently the case, as experience shows. Various species often occur together and mixed with one another at the spots from which the material for the observation was obtained ; during the investigation forms not desired, ' unbidden guests,' may find their way with particles of dust into the material, and even when we are dealing with apparently quite pure material, a small quantity of microscopic weeds, as we may say in this case also, may be mingled with it.

If every thing in the mixture grows at an equal rate, the different species may be kept distinct with comparative ease, and the character of the mixture is clearly understood. But the state of things may be different from this, and experience shows that it often is different. The one species develops vigorously under the existing conditions, the other feebly or not at all; the more successful species gains the upper hand of the less successful, dispossessing it and even entirely destroying it. Further examination shows that in some cases a weed has grown up in place of the wheat. This may very easily happen. We shall see further on that some Bacteria, for example, double the number of their cells under favourable conditions in less than an hour. Those to which the conditions are unfavourable may be seen, if a single specimen is watched continuously, to be dissolved and to disappear entirely in a few hours. By the combination of phenomena of this kind the character of any given mixture may be totally changed in a short space of time.

It is obvious that difficulties such as these do not invalidate our postulate, but on the contrary bring it out into sharper relief. Those who altogether deny the existence of species in Bacteria, with Billroth and Nägeli at their head, have in fact never undertaken a direct observation of continuity of development, and they are therefore not justified in denying their existence. Billroth has accurately examined and compared the forms, but has never continuously followed and checked the changes in a preparation or culture; the observation has been interrupted by intervals of sufficient length to allow of various things happening unobserved. Nägeli, as far as can be gathered from his publications, has not closely examined the forms at all, but grounds his conclusions, even when they are morphological, on non-morphological observations with respect to phenomena of decomposition on the great scale. One instance of this mode of dealing with the subject may be mentioned. Nägeli remarks that milk which has not been boiled turns sour when left standing for a time, but that boiled milk becomes bitter (9). He admits that the sourness is due to the presence of a Bacterium. He considers the bitterness to be the result of a change in the action of the same Bacterium caused by the boiling—a 'transformation of the definite ferment-nature of a single Fungus into a ferment of another kind.' Here it is assumed that a single Bacterium-species is present in the unboiled milk; the question is not asked whether there may not perhaps be several species in it, some of which predominate before, the others after the boiling, and whether the different changes in the milk may not be thus explained. But Hueppe's more recent researches have shown that such is really the true state of the case (10). Of the various Bacteria-forms present in the unboiled milk, Micrococcus lacticus is at first most active at a low temperature, and turns the milk sour by the formation of lactic acid; it is killed by boiling, but the spores of Bacillus Amylobacter, the Bacillus of butyric acid, which is also present in the milk retain their vitality, and this Bacillus causes the decompositions in boiled milk which give it a bitter taste.

Another instance of the same kind is the statement emanating from Nägeli's laboratory, that the hay-bacillus, Bacillus subtilis, is identical with Bacillus Anthracis, the Bacillus of anthrax. The two species are very like each other, and Buchner's observations certainly contain some true remarks about them, which will be discussed in Lecture XII. But the most striking characteristic of B. subtilis is the often-described germination of its spores, the growing out of the germ-cell from the transverse fissure of the spore-membrane at right angles to the longitudinal axis of the spore. The Bacillus of anthrax does not exhibit this phenomenon, as Buchner himself tells us. But due regard has nowhere been paid to these differences, so that it is still doubtful whether Buchner has examined B. subtilis at all. In this case too the morphological statement is without certain foundation and justification.

The increased attention bestowed by observers on this subject, beginning with the wheat-plant and going down through various larger forms of the lower plants to Bacteria, has done away one after another with the erroneous notions indicated, and led to the general adoption of the view above explained, that questions of species are essentially alike throughout the series of organisms. In the case of the Bacteria much still remains to be done; our knowledge of these is yet only in its infancy.

I say that increased attention is leading to this result. I should wish at the same time to point out once more the conditions which have been and which are of the first importance. As might be expected, the aids to investigation, apparatus, technical methods, reagents, &c., have been improved. To determine the questions which we are at present considering, minute organisms have to be isolated and unceasingly watched in order to see what proceeds from the single individual, if it developes. This end can only be obtained by means of cultures which can be followed with exactness under the microscope. A spore or rod in the preparation must be permanently fixed under the microscope, and the phenomena of its growth must be observed without interruption. This is done by help of the moist chamber, a contrivance in which the microscopic object protected from desiccation can be observed continuously under conditions favourable to vegetation. There are several varieties of apparatus of this description, which have their advantages and disadvantages according to the special case and also to the habits of the observer, but we must not enter into a detailed description of them here.

Fluids are usually employed as the medium in which the object is placed for microscopic observation and for culture, on account of their transparency. Living and especially moving objects readily change their position in a fluid and become mixed together. A method which greatly assists the fixing of an object where continuity of observation is required, consists in the use of a transparent medium which allows of the conditions necessary to vegetation, and is soft but not fluid, so that displacement of the objects and disturbance of the observation are more or less perfectly avoided. Such media are gelatine and similar substances, especially the gelatinous substance known in commerce as agar-agar and prepared from sea-weeds of the Indian and Chinese seas. Gelatine, as I understand, was first employed by Vittadini in 1852 in the culture of microscopic Fungi (11), and has been frequently used since that time, especially by Brefeld. Klebs more recently in 1873 (12) recommends it specially for the cultivation of Bacteria; cultures of these organisms have been conducted in recent times in a gelatinous substratum, especially by Koch.

Having thus glanced at the morphology and the history of the development of Bacteria, we have still to enquire what is their position in the organic world and their natural affinity to other organisms.   The question is of only secondary interest to us on the present occasion, and must not therefore be examined at any length.

If we compare the structure and development of Bacteria with those of other known creatures, as we must do to answer the above question, the arthrosporous Bacteria are seen to agree entirely in all essential points with the members of the plant-group of Nostocaceae in the wider sense of the word; only the Nostocaceae are furnished with chlorophyll in conjunction with another blue or violet colouring matter which is soluble in water, and are thus distinguished from the Bacteria which contain no chlorophyll.   There is no reason why the arthrosporous Bacteria should not be termed Nostocaceae which are devoid of chlorophyll.   Structure, growth, occasional formation of Zoogloeae, more or less constant motility, especially developed in the Oscillatorieae, a division of the Nostocaceae, are the same in the two groups, so that apart from the absence of chlorophyll there is no greater difference between them than between the separate species of either of the groups. This may be illustrated by the case of Leuconostoc described on page 22.   The name indicates that the plant entirely resembles in all respects the bluish-green species of the genus Nostoc which live in water and on moist soil, only it is colourless and white.   To this may be added, that most of the Nostocaceae attain to considerably larger dimensions than the Bacteria in their cells and in the aggregations of their cells, and that the members of the group which resemble the Bacteria are related to other forms of a more varied and higher differentiation and conformation.

The Bacteria which we have distinguished as endosporous entirely resemble the arthrosporous Bacteria in every respect except the peculiar formation of spores, and resemble no other known organisms.   We must therefore place them next to the arthrosporous division, at least for the present and in accordance with our present knowledge.

Hence the Bacteria have been arranged in one group with the Nostocaceae, and this group has received the name of Fission-plants or Schizophytes; the Nostocaceae which contain chlorophyll are Fission-algae, those which have no chlorophyll are Fission-fungi.

The entire group of the Schizophytes is somewhat isolated in the general system; a closer association with other groups cannot be established at present, and it would lead us too far away from our more immediate subject to enter further into the conjectures which may be formed about them. So much however is beyond doubt, that most Schizophytes, the Nostocaceae especially, have all the characteristics of simple plants. They show a very slight approximation to the Fungi, in the sense in which that term is used in the natural system, as has been already stated in the Introduction. We can only say therefore that the Bacteria, together with the rest of the Schizophytes, are a group of simple plants of a low order.

The old observers regarded them as belonging to the animal kingdom and to the group of Infusorial Animalcules, chiefly on the ground of their motility and in the absence of the basis required for a more exact comparison. At present there is no reason for separating them from the vegetable kingdom, though it is merely a matter of convention in the case of these simple organisms where and how we should draw the line between the vegetable and animal kingdoms.

V.

Origin and distribution of Bacteria.

WE commenced our survey of the mode of life of the Bacteria by explaining in what manner and from whence they make their way to the spots where we find them.

If we adhere to the general result of the foregoing considerations, namely, that Bacteria are like other vegetable growths,

we may at once assume that their origin is the same as that
of other plants, that is, that the Bacteria existing at any given
time have sprung from beginnings which proceeded from individuals of the same species, and experience shows that this
is really the case.   These beginnings may be spores or any
other cells capable of life; we shall here usually call them
germs.

The germs of living beings, especially plants, are extraordinarily numerous. They may be said to cover the surface
of the earth and the bottom of the waters with an infinite
profusion of mingled forms.   The number of plants observed
in the developed state gives no idea or only a very imperfect
idea of this fact, because a much larger number of germs is in
all cases produced from a single plant than can arrive at their
full development in the space at their command, which is in
fact always limited. The smaller the organisms are, the greater
advantages they enjoy as a general rule *caeteris paribus* for the
production and distribution of their germs, for it is so much
easier for them to find space and a sufficient quantity of food for
their development and for the production of new germs; the
mechanical conditions for the transport of the germs from place
to place are also more favourable in proportion as the volume
and mass are diminished.   For these reasons the number and
distribution of the germs of lower microscopic organisms, especially in the vegetable world, must seem astonishingly great to
any one who is unprepared for the facts.   If spring-water is
allowed to stand in a glass, it becomes green in time from the
growth of small Algae, whose germs were present in the water
before it was placed in the glass or have been carried there with
particles of dust.   If a small piece of moistened bread is placed
in the water a growth of mould soon makes its appearance, proceeding from germs of Mould-fungi.   Some time since I made
researches with a different object into the Saprolegnieae, a
group of rather large Fungi consisting of about two dozen
well-known species, which grow in water on the bodies of

dead animals, and it was found that germs of one or several species of this single group were present in every handful of mud from the bottom of every sheet of water from the sea-level to a height of 2000 metres. The actual presence of the germs may be shown in all these cases by microscopic and experimental examination, to the conduct of which we will recur presently.

As these facts again would lead us to expect, among microscopic growths some are rare and some are common, some have a limited and some a very extensive area of distribution. The principle must be the same with these as with the higher and larger organisms; climatic and other external causes must have a similar effect on the distribution, though for the reason stated above that effect is generally less powerful than in larger and more pretentious forms. The researches into this subject are not yet extensive enough to permit of the production of many details. But we know, for example, that a small Fungus, scarcely visible to the naked eye, Laboulbenia Muscae, which vegetates on the surface of the bodies of living house-flies in Vienna, and which appears to be common in southern Europe, does not occur in the middle and west of Europe; at all events after careful search it has not yet been found. Instances of the reverse kind are more numerous. Our common species of moulds, Penicillium glaucum, for example, and Eurotium, are spread over all parts of the world and all climates, and the same is the case with other small Fungi and Algae.

In this point also Bacteria are only special instances of the series of phenomena which have been shown above to occur as a rule in small organisms. Our knowledge of the several species, as appears from preceding lectures, is too imperfect to enable us to make precise statements with respect to the larger number of them; at the same time we know that some species are comparatively rare, such as Micrococcus prodigiosus and Bacillus Megaterium, while others, like B. subtilis, B. Amylobacter, and Micrococcus Ureae, occur in almost every situation

in which they find the conditions of vegetation, which are themselves of very common occurrence. We shall make acquaintance with other illustrative instances in subsequent special
discussions. Dispensing with an exact determination of the
species in every case, we shall be perfectly safe in declaring,
as the result of direct observations, that the vital germs of
Bacteria are scattered abroad with such profusion in earth, air,
dust, and water, that their appearance at all spots where they
find the conditions necessary for vegetation is more than
sufficiently explained.

The way to prove this, and at the same time to determine
approximatively the number of germs within a given space, is obviously the same in the case of the germs of Bacteria as in that
of other lower organisms, Fungi and others; both necessarily
come under our observation at the same time, when they are
present. It consists first of all and simply in microscopical
examination. But in this method we encounter considerable
difficulties. Sometimes the germs are not present in every
smallest spot; they must be sought for, and this is at all times a
troublesome process, especially when it is intended to count them.
Various devices may it is true be applied to lighten this labour.
Pasteur (13), for example, employed an ingenious contrivance
for finding germs in the air in the form of a suction-apparatus,
an aspirator, which drew in the air through a tube stopped with
a dense plug of gun-cotton. The plug allows the air to pass,
while the solid substances in suspension in the air and the
germs therefore with them are caught on or in the plug. The
quantity of air passing through the apparatus within a given
time can be easily determined. The gun-cotton is soluble in
ether, and by taking advantage of this property the germs which
have been intercepted in the plug may be obtained suspended
in a clear solution, and collected within a narrow space for examination and even for counting.

But in this process the germs are very liable to be killed by
the ether, and even in ordinary microscopical examination it is

impossible to be quite certain whether we are dealing with dead
or with living objects. Yet it is a matter of the first importance
to determine whether germs capable of development are present
or not, and this would require further and very complex modes
of procedure.

Hence various other methods have been tried with the object
of making the investigation easier and more trustworthy in both
directions. It was Koch who at length cracked the egg, like
Columbus. Starting from the empirical fact that gelatine, combined with other nutrient substances easily prepared and in a
state of solution, is a very favourable substratum for the development of most Fungi that are not strictly parasitic, and also
of Bacteria, he distributes quantities of the substances intended
for examination, earth, fluids, &c., in properly prepared gelatine,
liquefying at a temperature of about 30° C., and then makes the
gelatine stiffen by lowering the temperature. The quantities
may be exactly determined. Each germ is fixed in the
stiffened mass and so developes, and the products of the
development are at least at first also fixed and not liable to
displacement in the medium. If the transparent gelatine is
spread in a thin layer on glass slides at the commencement of
the investigation, the germs and the products of their development can be found with certainty with the microscope, and if
necessary be counted. If the object is to examine the air, the
best plan is to draw it in slowly by means of an aspirator
through glass tubes, coated inside with a layer of gelatine. If
the stream is properly regulated, the greater part at least of the
germs which are mixed with the air sink downwards and are
caught in the gelatine, where they may then undergo further
development. If experiments of this kind are properly conducted and disturbing impurities excluded, distinct groups of
Bacteria, Fungi, &c., will be found after a few days in the
gelatine. Each group originates in a germ, or in some cases in
an assemblage of germs, which made its way to the particular
spot at the commencement of the experiment, as may often

be easily ascertained by direct observation. It is obvious that
the purpose under consideration can be most certainly and most
simply effected in the way which has just been described. The
result certainly can never be more than approximatively exact,
because the nature of the process does not ensure that all the
germs capable of development which find their way to the
gelatine in the apparatus do in any given case actually develope,
or in the case of air-suction that all the germs without exception
are always actually caught. No other method which has not
this fault in an equal or even greater degree and without the
advantage of fixing the germ has up to the present time been
devised, nor is it easy to imagine one that would be practicable.
It may be added here that Koch's method has the further
advantage of making the sorting and selection of Bacteria for
isolated culture comparatively easy. Each of the groups derived
from a single germ in the experiments above described must
contain a single species without admixture. To obtain a
quantity of this species for a pure culture we have only to remove
a sample from the group with the needle. To sort a mixed mass
of Bacteria requires simply the spreading small quantities of it
over a large amount of gelatine, and thus isolating germs capable
of development. The groups formed from these germs supply
pure species-material. Various other experiments have been
tried with the same objects and on the same principles, but with
less perfect arrangements and methods; we must not, however,
enter here into a more detailed account of them. The most
elaborate are those instituted by Miquel and continued from
year to year in the meteorological observatory at Montsouris,
near Paris, intended especially to ascertain the distribution of
germs in the air and in water (15).

All researches hitherto conducted have given the general
result described above, and a further one which might have
been expected beforehand, namely, that the number of germs
capable of development varies, other conditions being the same,
with the place, the time of year, the weather, and other circum-

stances. To give some idea of the approximate numbers it may be added, that the number of germs in the air caught on glass plates in a mixture of glycerine and grape-sugar in the aspirator, Fungi and Bacteria capable of development and in some cases dead being taken together, varied in the garden of Montsouris, in a single series of observations, from between 0·7 to 3·9 in December and to 43·3 in July in a litre of air.

The most exact air-determinations have been recently carried out by Hesse with the aspirator and gelatine-process. These showed the presence of germs capable of development in a litre of air, as follows:

In Sick-ward No. 1, with 17 beds, Bacteria 2·40,—Moulds 0·4.

      „   „   2, „ 18 „   „   11·0,   „   1·0.

Cattle-stall for experimental pur-

    poses belonging to the Na-

    tional Office of Health: (*a*)   „   58·0,   „   3·0.

                               (*b*)   „   232·0,   „   28·0.

The air out of doors in Berlin was found to contain 0·1—0·5 germs per litre, of which about half were Fungi and half Bacteria.

Miquel obtained thirty-five germs per cubic centimetre in rain-water caught as it fell, sixty-two in river-water from the Vanne; in that from the Seine above Paris 1400, below Paris 3200.

We have no numerical determinations of the number of germs present in the soil; but we can produce growths of Fungi and Bacteria from every small pinch of soil taken from the surface of the ground. In lower strata, according to some preliminary researches made by Koch (14) in winter, the number of germs capable of development diminishes rapidly.

A special interest attaches to the question of the presence of germs in and on sound living organisms. That they must remain hanging in profusion to the surface of such organisms is obvious from the preceding statements, and is proved by every investigation. They can penetrate into the interior of the higher forms of plants through the open slits in the epidermis, the stomata, which lead to the system of intercellular passages. It is probable that this actually takes place, but it is not yet quite certain and requires further investigation. The respiratory and alimentary canals in healthy, especially warm-blooded, animals are constantly accessible places for the entrance of germs with air, meat and drink, and it is these parts, especially the mouth and the intestinal canal, both in man and other warm-blooded animals, which are as a matter of fact always a well-stocked garden of vegetating Bacteria. They may also make their way into the glands which are in communication with these canals through their excretory ducts. Researches into their occurrence in the blood of healthy living warm-blooded animals give different results. Hensen, Billroth, and other observers maintain their presence there. Very careful investigations by Pasteur, Meissner (13), Koch, Zahn, and others give a negative result; the affirmative result may therefore be due to disturbances and errors in the experiment. But this conclusion is not unavoidable, for a series of experiments by Klebs (12) have placed it beyond doubt that both states may occur, and why they may occur. Klebs examined the blood of some dogs, and partly with a negative result. But in the case of one dog the result was affirmative, the fact being that putrefactive Bacteria had been injected into the blood of this animal some time before on the occasion of some other experiments; it had sickened with them but had quite recovered long before the date of the investigation of which we are now speaking. It cannot be doubted that in this case germs capable of development but actually dormant had remained from the first experiment in the animal's blood, and we may conclude generally that Bacteria germs may be present in healthy blood, if they have once made their way into it through a wound or in some other way.

The result of the above facts is to show the wide distribution and great abundance of Bacteria-germs, though their species are not at present clearly discriminated. They show also on the other hand that it would be an exaggeration to suppose that these bodies are everywhere present, that is, in every minutest space. Even Pasteur's earlier and famous researches show the inequality of the distribution by extreme examples. This may be briefly illustrated by the following account. A small quantity of germ-free nutrient fluid, very favourable for the development of lower organisms, was introduced into small narrow-necked phials of 1–200 ccm. content; the air was withdrawn from the phials and the narrow neck hermetically closed. Subsequently the closed neck was reopened by intentional fracture of its extremity; air rapidly poured in, and as soon as this had taken place the neck was once more closed. From 1–200 ccm. of air were thus hermetically inclosed in the phial. The germs which they contained were at liberty to develope in the nutrient fluid, which, to use a short expression, remains unaltered if no germs are present. Of ten such phials filled with air in the court-yard of the Paris Observatory not one remained unaltered; nine out of ten filled in the cellar of the Observatory, which was almost entirely free from dust, and nineteen out of twenty filled at the Montanvert near Chamonix were unaltered.

The views here expressed with regard to the origin of Bacteria, and especially the fundamental axiom, that they are produced without exception from germs springing from species of the same name, have not been arrived at without trouble or without opposition, and the latter has not entirely ceased even at the present day. We must not pass by the view of the opponents without at least a brief consideration. It may be concisely stated thus: Bacteria may be formed at any moment from parts of other organisms, living or dead; but it is allowed that they may afterwards multiply by their own growth and also produce their own germs.

This view is a survival from the old doctrine of original production without parents, spontaneous or equivocal generation. Plants or animals are often known to appear in numbers in places where they had never been seen before, and the superficial observer is led to assume in such cases that they owe their origin to other bodies present at the particular place before their appearance there, no matter what these bodies may be, and not to germs formed from similar parents. Such views were not unnatural in ancient times. Virgil's (16) account of the production of a swarm of bees from the buried entrails of a steer furnishes an obvious illustration, and shows how utterly defective were the observation and reasoning which admitted of such notions. With a closer observation of nature it became evident in one case after another that the appearance of the particular organisms invariably commenced with germs which were the product of parents of the same kind, and that the point not observed was how these germs found their way to the place of observation. Generation without parents was step by step driven into a corner. The process began with large and coarse objects like the maggots of flies which appear in carrion, not by spontaneous generation, but produced from the ova of flies which have been deposited in it. And as the adherents of the old doctrine were driven back on smaller objects, such as moulds, the lowest forms of animal life and the like, their refutation followed step by step with equal success in these domains also. Microscopic and improved experimental methods by turns sharpened the weapons. Thus we find ourselves face to face with the fact that the adherents of generation without parents, at least during the last hundred years, seek for support to their doctrine always in the minutest and at the time the most inaccessible objects. The view has never been entirely given up, and for two good reasons. First, because an opinion once expressed or put into print, be it what it may, never totally disappears ; the second and much better reason is, that we must necessarily assume that organisms were certainly once produced without germs and without parents ; the possibility that this may happen again at any time must be allowed, and to prove that this does happen and to show where and how it happens would be highly interesting, and a really worthy subject for the efforts of the enquirer.

Bacteria rank with the smallest organisms at present known to us, with the least accessible and the most imperfectly investigated. It is true that the question of actual spontaneous generation has been in all essential points decided in the same way in their case as in that of other organisms, by the beautiful researches conducted by Pasteur twenty-five years ago at the instance of the Academy of Paris, and intended to test the doctrine in question in connection with the smallest and least accessible creatures; and every pure and trustworthy investigation has confirmed Pasteur's results. Nevertheless there are writers who still hold to the doctrine and who seek for fresh arguments in support of it. A comprehensive theory in this direction is contained in Béchamp's doctrine of Microzymes (17) published twenty years ago. The term Microzyme was applied by him to minute form-elements, such as occur generally in the shape of granules in the protoplasm of animals and plants, and are doubtless formed in them as parts of their substance. If these particles of matter are set free by any cause, especially after the death of the parent, they are supposed to undergo a further process of independent development and to become partly Bacteria, partly also small Sprouting Fungi. They not only outlive the organism which produces them, but enjoy a very prolonged existence extending over geological periods. Close scrutiny of the accounts given by Béchamp in a volume of almost a thousand pages shows no sharp discrimination of forms, and no sign that the continuity of the development has been strictly followed, and yet this is a point of the very first importance. The whole matter therefore is without any certain foundation, and is no longer a subject for discussion.

A. Wigand (18) has quite recently published a preliminary communication, in which he arrives at the same results as Béchamp as regards the question before us. Small portions of living or dead organisms, the latter not being Bacteria, are said to separate from them under definite conditions and to develope into Bacteria. The course of the observations, from which this conclusion is drawn, is in most cases not stated with sufficient exactness to allow of our forming a judgment upon them. Still one observation is mentioned which it was admissible and desirable to have repeated and tested. Wigand states, for the removal 'of all doubt about spontaneous formation of Bacteria in the protoplasm of cells,' that motile Bacteria are found in the living healthy cells of the leaf of Trianea bogotensis and in those of the hairs of Labiatae. My attention had been directed to the matter from another quarter before I proceeded to examine into this remarkable statement. Trianea is a South American water-plant, which floats in the manner of our Frogbit (Hydrocharis). If living tissue from the fresh healthy leaf is placed under the microscope, we shall really see in many cells the prettiest representations of the appearance of Bacteria, small slender rods, isolated or attached together in short rows and actively following the movements of the protoplasm and other cell-contents. An excellent representation, as I said, or model. But a drop of dilute muriatic acid destroys the illusion. The acid at once dissolves the rods in Trianea, which it would not do if they were really Bacteria; they are simply small crystals of calcium oxalate, which often occur in vegetable cells and in the form of rods. Of the same kind are the much less beautiful rods in the young hairs of the leaf of Galeobdolon luteum and Salvia glutinosa, and so also in other Labiatae or lipped-flowered plants. The case is full of instruction, as showing how a preconceived opinion may lead even good and intelligent observers into the greatest absurdities. I should not otherwise have mentioned it, and I do not think it necessary to go any further into similar matters. Such things at all events are not calculated to weaken the proposition, that according to the observations which actually lie before us even the smallest organisms spring only from germs produced from ancestors of the same kind; and to this we must hold fast in spite of whatever may be thought possible or desirable.

VI.

Vegetative processes. External conditions : temperature and material character of the environment. Practical application of these in cultures, in disinfection, and in antisepsis.

IN passing on to the consideration of processes of vegetation, we must first of all remember that agreement in structure and development between Bacteria and other lower organisms necessarily implies also an agreement in the chief phenomena and chief conditions of vegetative life. In fact we have simply to do with special cases of phenomena which are of general occurrence in all living organisms, and which do not differ more from those to be met with in other plants than these do from one another. It is specially true of the Bacteria which do not contain chlorophyll, that their vegetative process agrees essentially with that of other vegetable cells which do not contain chlorophyll, both those which belong to the higher plants, and more particularly those belonging to the Fungi. It is to the investigation of the Fungi, which are more easily studied, that we owe much of the advance that has been made in our knowledge of the Bacteria. It is perhaps scarcely necessary to observe that differences prevail from one case to another among Bacteria also as regards the phenomena and conditions of vegetation, analogous with those in the allied groups.

Our present object, however, is not to give a complete account of everything belonging to the vegetative process, but only to call attention to the points most worthy of notice in connection with the subject of these lectures. The conditions of temperature and the material character of the environment are chiefly to be considered.

Every process of vegetation is dependent on the temperature of the surrounding medium ; it finds its limits within certain extreme degrees of heat, and its greatest activity at a fixed

E

temperature between these extremes. The cardinal points of temperature are accordingly distinguished as minimum, maximum, and optimum.

Transgression of the limits leads at first to a cessation of the particular process going on at the time; other processes may possibly persist. If the raising or lowering of the temperature beyond the maximum or minimum point of vegetation reaches certain extreme degrees, life is destroyed, in other words the death-point is attained.

In all these respects considerable variations occur in conformity with every one's daily experience, according to the species, the state of development, and the character of the environment.

The limits of temperature in the growth and multiplication of cells are the points which have been chiefly examined in the case of the Bacteria; it being assumed with some reason that the rest of the vegetative processes, other conditions remaining the same, run proportionally with the growth.

It appears from the data before us that non-parasitic species, if well and properly nourished, have a tolerably wide range and a high optimum of growth-temperature. The former lies in Bacillus subtilis, for example, according to Brefeld (19), between 6° C. and 50° C., the optimum being at about 30° C. Bacterium Termo, Cohn grows between 5° C. and 40° C., while its optimum is 30–35° C. (Eidam 20). Bacillus Amylobacter, according to Fitz (21), has its optimum in solution of glycerine at 40° C., its maximum at 45° C. The minimum of growth, according to present accounts, in Bacillus Anthracis in cultures in gelatine, on potatoes, &c., is at 15° C., the maximum at 43° C., the optimum at 20–25° C. As a parasite in the blood of rodents it grows at about 4° C.; at least, not less vigorously than in the optimum just given for specimens under culture. In the Spirillum of Asiatic cholera, according to van Ermengen (see Lecture XIII), the minimum is reached at 8° C., the optimum at 37° C., the maximum at 40° C.

That the species which are more strictly adapted for a para-

sitic life in warm-blooded animals have a higher maximum and
optimum is probable beforehand, and has been proved by Koch
(60) in the case of the Bacillus of tubercle, in which the limits
of temperature were found to be from 28° to 42° C., and its
optimum 37–38° C.

The optimum temperature for the formation of spores in
endosporous Bacilli, as far as can be ascertained, approaches
that of growth. The temperatures for the germination of the
endogenetic spores are higher, at least in the case of the optimum, being 30–34° C., for instance, in Bacillus subtilis, which
however also germinates in the temperature of a room which
is somewhere about 20° C. B. Anthracis does not germinate,
as far as our experience goes, at 20° C.; the minimum given
for this species is 35–37° C., the optimum can scarcely be much
higher. Other species, as B. Megaterium, grow and germinate
quite well at a temperature of about 20° C.

Transgression of the limits of temperature of vegetation in
the downward direction without destruction to life is possible in
the case at least of a large number of Bacteria, and to such an
extent that in view of the phenomena which are known to occur
we may even say that there are no limits. Frisch (22) found
the power of development in the forms which he examined, and
in their vegetative cells, unaffected when they were frozen in a
fluid at a temperature of −110° C. and afterwards thawed
again. Bacillus Anthracis is one of the forms which behave in
this manner; in the case of other species the point remains
undecided, but it is probable that in some of them the lower
death-temperature is higher than this.

The upper death-temperature, so far as is at present known,
is about the same for the vegetative cells of the majority of
forms, as for most other vegetable cells, 50–60° C. Similar
figures are true also for the spores of arthrosporous forms,
though this point requires further investigation. Exceptional
cases will be mentioned further on. On the other hand, the
endogenetic spores of the Bacilli are capable of enduring

extreme high temperatures. Most of them continue capable of germination after being heated in a fluid up to 100° C.; some will bear 105° C., 110° C., and as much as 130° C.

These are all general rules and are not affected by the modifications and exceptions which occur in different cases, and which in part depend on the species and individual, other conditions remaining the same, in part also are found in the same species, being then dependent on the external conditions, such especially as the length of the time during which they are heated, dried, or soaked, and in the latter case on the nature of the surrounding fluid.

There are first of all species which develope vigorously at a temperature considerably above 50° C. Cohn and Miquel supply instances of this, but the best is that of a Bacillus described by van Tieghem (23), which grows and forms spores in a neutral nutrient solution at a temperature of 74°C.; growth ceases at 77°C.

The Bacilli obtained by Duclaux (24, 25) from cheese, and named by him Tyrothrix, are instructive examples on all the points above-mentioned. The vegetative cells of T. tenuis cultivated in a neutral fluid were only killed at a temperature of 90–95° C., in a slightly alkaline fluid they bear a temperature of over 100° C., while the ripe spores remain capable of germination when subjected to a temperature of 115° C. in a similar fluid. The most favourable temperature for vegetation in this species is 25–35° C. T. filiformis in the vegetative state will bear a temperature of 100° C. in milk, but is killed in the space of a minute in an acid fluid at the same temperature. The spores of this species are uninjured at a temperature of 120°C. in milk, but are killed at less than 110° C. in gelatine. Duclaux gives similar accounts of other species. The vegetative cells also of Bacillus Anthracis are said by Buchner (28, p. 229) to continue capable of infection when heated for an hour and a half in neutral and slightly acid fluids up to a temperature of 75–80° C. Brefeld (19) found all the spores of Bacillus subtilis in a nutrient solution kept for a quarter of an hour at a temperature of

100° C. capable of germination; if they remained in it at the same temperature for half an hour the majority still germinated, if for one hour a smaller number; none retained their vital power after a space of three hours.   The spores were killed in fifteen minutes at a temperature of 105° C., in ten minutes at 107° C., in five minutes at 110° C.

Fitz (21) found that the spores of his Bacillus butylicus (B. Amylobacter) bear a temperature of 100° C. for a time varying from three to twenty minutes, according to the fluid in which they happen to be.   But if the time of exposure is prolonged, temperatures under 100° C. are sufficient to kill them, 80° C. for example, when they are kept seven to eleven hours in glycerine solution.

Spores, at least, are proof against still higher degrees of dry heat; those of Bacillus Anthracis, B. subtilis, and others continued capable of development in Koch's experiments (14, p. 305) in a chamber heated up to 123° C.

Among the conditions connected with the nature of the environment, the requisite supply of water must be mentioned first in this case as in that of all living cells.   Withdrawal of water to the point of air-dryness not only stops the process of vegetation but kills vegetative cells, at least in a number of cases, in a very short time, those of Bacterium Termo, Cohn, and B. Zopfii, for example, in seven days.   But here, too, the effect varies in different cases; Micrococcus prodigiosus, for instance, continues alive and capable of development for months in a state of desiccation.

The resistance of spores to desiccation is greater than that of vegetative cells.   The spores of the arthrosporous Bacterium Zopfii withstand it for seventeen to twenty-six days; those of the endosporous Bacilli on the average certainly a year, those of Bacillus subtilis, according to Brefeld, at least three years. Here, too, limits and modifications will arise according to other internal and external causes, but air-dry cells can hardly be expected to retain their vitality for centuries.

Oxygen is not equally necessary in all cases. Two extreme cases are distinguished in Pasteur's terminology as aerobia and anaerobia. The first require an abundance of air containing oxygen, as well as a good supply of nutrient substances for luxuriant vegetation and growth; of this kind are Micrococcus aceti, Bacillus subtilis, B. Anthracis, and Koch's Spirillum of cholera. The other kind does well on good food without oxygen; free access of air reduces their vegetation to a minimum or to zero, as for example in Bacillus Amylobacter.

Intermediate cases, however, are found between the two extremes, as is well shown by Engelmann's beautiful example which will be referred to again presently; and according to the investigations of Nencki, Nägeli, and others, Bacteria which excite fermentation, like the Sprouting Fungi which give rise to alcoholic fermentation, grow luxuriantly without oxygen, when they are in a suitable fluid capable of fermentation with them. If these forms are placed in a less favourable nutrient fluid in which they cannot incite fermentation, they will not grow without a supply of oxygen.

Oxygen may impede and even destroy vegetation even in the case of aerobiotic forms if it takes place under high pressure. Bacillus Anthracis, for example, remained alive for fourteen days in oxygen under a pressure of fifteen atmospheres, but was dead in a few months' time. Duclaux contends that the germs even of aerobiotic forms, when withdrawn from the conditions required for growth, lose their power of development more quickly under the continued effect of atmospheric oxygen than when oxygen is excluded. The facts on which this view is founded are in themselves remarkable. In some glass bottles which had been used in Pasteur's researches about 1860, and had been kept hermetically sealed with their contents decomposed by Bacteria, the germs of these Bacteria were found thoroughly capable of development after twenty-one and twenty-two years. Plugs of cotton-wool full of germs of all kinds, which had been kept dry and protected from dust during the same time, but not from contact with the air, did not contain a single living germ. A few similar plugs which were only six years old contained germs still capable of development. Duclaux' interpretation of these facts may be correct, but it requires further proof, since we are dealing with matters in which many other things besides the supply of oxygen may have been unequal. Above all things it is necessary in these questions that experiment should be made, not with collective Bacteria, that is, with mixed masses which are possibly or certainly undetermined, but always with a single definite species.

Oxygen is taken up as material for respiration or breathing, oxygen-breathing, to use a more precise term, carbon-dioxide being at the same time given off. Water, except in some cases which will be mentioned presently, serves as the agent and medium of the chemical processes of the metabolism. Neither of these bodies is properly a nutrient substance, that is, a substance from which carbon-compounds, the constructive material for growth and cell-formation, are produced.

With respect to the true nutrient substances which therefore supply building-material we must assume in the case of the few green Bacteria, if they really contain chlorophyll, that according to the analogy of all other plants containing chlorophyll, they assimilate carbon as their food and give off oxygen. Engelmann (26) has ascertained that a small portion of oxygen is given off by his Bacterium chlorinum, and this supports the assumption, while the employment of water also as a food-material in the case of these forms, as in all other plants containing chlorophyll, would also be probable.

The Bacteria containing no chlorophyll, which are by far the greater number and almost the only ones which concern us at present, require, like all cells and organisms that are devoid of chlorophyll, carbon-compounds previously formed else-where for the supply of their carbon, and do not assimilate carbon-dioxide. The nitrogenous food-material may be furnished both by previously formed organic and also by inorganic sub-

stances, compounds of nitric acid or still better of ammonia.
In addition to these a small supply quantitatively and qualitatively is required, as in other plants, of soluble constituents of
the ash.

It does not fall within the scope of these lectures to go more
deeply into the consideration of the value of the several compounds as food-material; on this point the special literature,
especially Nägeli's publications (27, 28), should be consulted.
It is sufficient for our general guidance and for practical purposes to observe that according to Nägeli's investigations a
number of moulds and Sprouting Fungi as well as Bacteria also
can find their food in solutions which contain nitrogenous and
non-nitrogenous nutrient substances in the following compounds
or combinations, the several solutions being arranged and
numbered in descending order according to their nutritiveness:—
1. Proteid (peptone) and sugar. 2. Leucin and sugar. 3.
Ammonium tartrate or sal-ammoniac and sugar. 4. Proteid
(peptone). 5. Leucin. 6. Ammonium tartrate or ammonium
succinate, or asparagin. 7. Ammonium acetate.

But we must not seek to determine or judge of the optimum
of feeding - quality for all species or forms of Bacteria from
this table. The above scale is not even true for all moulds,
though it was first drawn up from the study of one of that group,
Penicillium glaucum. The requirements in the way of food of
single definite species of Bacterium have as yet been little studied,
and much needs more exact investigation. A number of practical experiences, which will be partly noticed further on under
the particular examples, point already to the great multiplicity
of the actual relationships which have to be taken into account.

Besides the amount of suitable food-material contained in the
substratum, other chemical qualities in it are also of importance
to the vegetative process in Bacteria. It is an old experience
that most of these organisms, in contrast to the reverse behaviour of Sprouting Fungi and moulds, flourish best, other
conditions being the same, in a medium with a neutral or

slightly alkaline or at most with a slightly acid reaction ; should
the reaction be strongly acid, vegetative processes are hindered
or wholly stopped.   According to Brefeld (19) the development
of Bacillus subtilis, for example, is impeded, if o·o5 per cent. of
sulphuric or tartaric acid or o·2 per cent. of lactic or butyric
acid is added to a good nutrient solution.   But this, too, is
only a rule which has its exceptions ; the Bacterium of kefir
vegetates well, and, as far as our experience goes, best in milk
which has been rendered strongly acid by lactic and even acetic
acid ; the Micrococcus of vinegar vegetates in the same way in
an acid fluid.

Other soluble bodies also impede or destroy the vegetative
process when mixed with the food-material.   This is of course
the case with substances which always act as poisons upon living
cells, such as corrosive sublimate, iodine, &c., when present in
sufficient quantity.   But other bodies have a similar at least re-
tarding poisonous effect on Bacteria.   Fitz, for instance, found
that the vegetation of his Bacillus of butyl-alcohol in a solution
of glycerine and under conditions otherwise most favourable was
impeded by the addition of 2·7–3·3 per cent. by weight of ethyl-
alcohol, o·9–1·o5 per cent. of butyl-alcohol, or o·1 per cent. of
butyric acid.   Since these prejudicial compounds are often
formed by the vegetative process itself, the latter may even be
stopped by the accumulation of its own products, as, for instance,
in lactic acid fermentation in sugars by the accumulation of
lactic acid ; if this is fixed, as by addition of chalk or zinc-
white, the vegetation of the Bacterium which causes the fermentation continues.   These phenomena are also found *mutatis
mutandis*, in other plants beside Bacteria, especially in Fungi,
and they vary in the individuals of different species.   That
which disturbs one species may be of advantage to others, and
hence a change in the composition of the substratum may
favour the supplanting of one species by another, which was
previously perhaps present in the very smallest quantity.   In
such a case the first species has prepared the ground for the

others by its vegetative process and its products. This must always be kept in mind in judging of processes on the large scale; attention to it supplies the explanation of a number of phenomena which are at first sight puzzling.

The influence of other agencies besides those which have been mentioned on the vegetation of Bacteria cannot in general be disputed, but in the present state of our knowledge it is of so subordinate importance, that a very short notice of it will be sufficient on this occasion. The dependence of carbon-assimilation upon the rays of light in the forms which contain chlorophyll follows of necessity from what we know of the function of chlorophyll. With respect to other effects of light we have only some uncertain statements by Zopf on the probable promotion of the growth of Beggiatoa roseo-persicina by illumination, and an investigation by Engelmann (29) into the dependence on the rays of light of the movements of a form which, though named Bacterium photometricum, is possibly, to judge by the illustrations, not a Bacterium at all. Influence of light has not been proved in the case of the majority of Bacteria. The effects of electricity have been recently investigated by Cohn and Mendelssohn (30), and may be gathered from their paper.

The dependence on the conditions of vegetation which we have been considering is true of all stages and phases of the normal vegetative process, not excepting its first beginnings, the germination of the spores. Of this it must be specially remarked that it occurs, as far as is at present known, only in a nutrient substratum favourable to the vegetation of the species. This agrees with the corresponding behaviour of some spores of Fungi, those for example of Mucorini. It does not agree with that of most other spores or with the seeds of flowering plants, which germinate, or at least can germinate, without nutrient substances, provided they are supplied with water, oxygen, and the necessary warmth.

It has been already stated above on page 19, that in some cases, as in Bacillus Amylobacter, spore-formation takes place even while vegetation and growth are going on in a portion of the vegetative cells, and therefore while the conditions of vegetation are still in operation. In other and especially in the endosporous species it is true to say, that the formation of spores begins when the substratum is exhausted, that is, has become unsuitable for the vegetation of the species. Whether the latter condition is really due in every case to a consumption of the requisite nutrient substances or to an accumulation of checking products of decomposition, or whether the formation of spores is induced in this case as in others by internal causes when the vegetation has reached a definite height, are all questions which require more precise investigation, though they may perhaps be of only subordinate practical importance.

Vegetation proceeds with great rapidity in most Bacteria under the co-operation of the most favourable conditions. Brefeld determined in the case of Bacillus subtilis, that with a good supply of food and oxygen, and a temperature of 30° C., a rod divides once in every thirty minutes, which means that it doubles its length every thirty minutes, the thickness remaining the same, and then separates transversely into two equal parts. The process goes on more slowly in proportion as the conditions recede from the optimum. If we assume that the increase directly observed in the way here described is accompanied by a corresponding increase in the mass, especially of the dry substance, an assumption which is not strictly proved but from the indications before us is certainly approximatively correct, then we have growth to double the former size in the full sense of the expression once in every thirty minutes. Similar results are arrived at from observations on many other species, as Bacillus Anthracis, B. Megaterium, &c. But here, too, there are exceptions. The Bacterium of kefir, for example, in the cases which I examined, required more than three weeks for growing to about twice its weight, more than 500 times the period observed in Bacillus subtilis. I am not able to say whether the conditions were absolutely the most favourable; at all events, they were those in which the kefir-organism grows best according to our present knowledge, namely, in milk at an air-temperature of 15–20° C., and with a supply of atmospheric air.

The movements also of the Bacteria, as well as their growth and germination, are directly dependent on the conditions of vegetation in the species and forms which are capable of independent movement. The occurrence and the direction of the motion are specially determined by the influence of nutrient substances, and of oxygen. If a form of this kind, Bacillus subtilis for instance, in the vegetative condition in which it is capable of movement is placed in a drop of nutrient solution on a slide under a cover-glass, the motile rods are seen to collect at once round the margin of the cover-glass where the oxygen of the air has free access. The comparatively few which remain behind in the centre of the drop, and are there cut off from the atmospheric oxygen, become slower in their movements and finally lose them altogether. Aerobiotic forms enclosed in a drop of water in which there is no free oxygen along with Algae containing chlorophyll at first remain motionless. But as soon as the cells containing chlorophyll are induced to give off oxygen under the influence of light, the Bacteria begin to move actively, as Engelmann (31) has shown, and the movement is directed towards the spots where the oxygen is being given off. Here the Bacteria collect, and they may therefore be used as an extremely delicate reagent for the detection of quantities of oxygen of almost inconceivable minuteness. The frequent grouping of aerobiotic forms into films or membranes on the surface of fluids is no doubt partly due to the influence in question determining the direction of the movement.

While the above-mentioned forms approach as near as possible to the source of the atmospheric oxygen, there are others which, as Engelmann (26) found in the case of a Spirillum, always remain at a certain distance from it, the distance diminishing as the amount of free oxygen diminishes in the air which finds access to the Bacteria. This observation proves the existence of intermediate cases, mentioned above, between extreme aerobia and anaerobia.

Pfeffer (32) has further shown that chemical stimuli, exerted by other bodies in a state of solution, may influence cells which have the power of locomotion and organisms of very various kinds, hastening and determining the direction of their movement, and that the Bacteria supply special instances of this general phenomenon. The chemical bodies which have this effect on the Bacteria are those which were spoken of before as their nutrient substances. The direction of the movement is due, as Pfeffer shows, to diffusion-currents by the introduction of the solutions on one side, the axis of rotation of the cells being in the same direction as the currents and the movement in space in the opposite direction. Other conditions remaining the same the effect varies according to the quality of the body in solution and the concentration of the solution, and it must be particularly observed that it is not every diffusion-current that influences the direction of movement, but only the current from solutions determined in each case by the species of Bacterium. These facts explain a phenomenon which has been frequently observed, namely, that swarms of Bacteria assemble in water round solid bodies, such as dead parts of plants, pieces of flesh and the like, which gradually give off soluble nutrient substances.

The practical application of these remarks on the conditions and phenomena of vegetation in conjunction with the ascertained facts respecting germs and their dissemination are in the main obvious, if the important points and conditions in each case are kept clearly in mind. We require always a certain amount of positive knowledge and careful consideration of the object which we desire to attain and can really attain in a particular way. The practical remarks therefore may, for the present, be summed up in a very few words.

First, with respect to the culture of Bacteria, there is but little to be said. Pure extracts of animal and plant-substances, the meat-extracts sold in the shops, broths, the juice of

fruits, neutralised, if necessary, and dissolved in not too concentrated (about 10 per cent.) watery solutions or in gelatine,
are, as a rule and in accordance with general experience, good
nutrient substrata; the special choice must be made by experiment in each case. Fresh urine has been repeatedly used with
success by French observers. The serum of blood has been
found to be a very suitable substance, and is almost the only
one that can be used in the cultivation of some parasitic forms,
especially if made stiff by being heated up to 60–70° C. after the
mode of proceeding described by Koch.

Among the very first requisites are the securing the purity of
the species under cultivation, the absence of unintentional admixtures, on which point some practical hints were given in a
former lecture (pp. 34 and 41), and the perfect control of the
continued purity of the cultures. The possibility of different
species displacing each other has been already discussed (p. 32).

To obtain purity of a culture as well as for other practical
purposes, it is often necessary to effect the entire destruction or
death of germs present in it. In the conduct of cultures there
is the special risk of these germs adhering to the apparatus to
be employed, vessels, nutrient substances, &c., and they must
be killed in order to provide for the purity of the culture. This
process of destruction is known as sterilisation, an expression
introduced by the school of Pasteur.

Bodies poisonous to protoplasm, such as acids, corrosive sublimate, &c., if sufficiently concentrated will usually effect the desired result, where the object is only to destroy, of course on the
one condition that they are able to force their way into the protoplasm which is to be killed. This is the case in most poisons
but not in all. Absolute alcohol is a poison which is immediately fatal to protoplasm, and it must therefore kill the protoplasm of endosporous Bacilli, if it reaches them. Nevertheless,
the spores of Bacillus Anthracis, as Pasteur discovered, and no
doubt also those of other endosporous species retain their vitality
after lying several weeks in absolute alcohol. If the same ex-

periment is made with sound ripe seeds of the ordinary garden cress, Lepidium sativum, the same result is obtained; they germinate if they are taken from the alcohol after four weeks' time, and washed and sown. The spores of the Bacillus and the seeds of the cress agree in being enveloped all round in a gelatinous membrane into which the alcohol cannot penetrate, and thus the protoplasm, which in the cress-germ would otherwise be certainly killed at once, remains unattacked.

But the application of poisons to cultures for purposes of sterilisation is attended with great inconveniences in all the many cases in which they must be got rid of again that they may do no injury to the culture itself. New impurities may be introduced in the process of washing the vessels and the rest of the apparatus.

Hence much the most practical mode of sterilisation consists in the application of extremely high temperatures, which must exceed 100° C., if the object is to kill any spores that may possibly be present; in dry vessels it is best to raise it to 120–150° C. In the sterilising of fluids a heat of even 100° C. may not always be possible for practical reasons, as, for example, when it is necessary to avoid the coagulation of the albuminous substances dissolved in the fluid. Since most vegetating cells are killed by a temperature of 50–60° C., the plan suggested by Tyndall (33) is the most effective ; the fluid is allowed to stand till whatever germs it contains begin to grow; if it is then heated to 60–70° C. and the process repeated at an interval of two days, the fluid will be in most cases free from Bacteria, always presupposing of course that the plug which closes the vessel is compact and clean.

Lastly, in practical life all that is usually required is to render harmless any germs that may be present by preventing their further development, whether they continue capable of it or not. Here, too, complete destruction would be best and most desirable ; but the use of most poisons in the state of concentration which is most certainly fatal, or that of a certainly fatal degree of heat, would also ordinarily lead to the destruction of the objects intended to be protected from the Bacteria. We must therefore be content with what is within our reach.

If, as there is no reason to doubt, the favourable results of the application of disinfectants at the present day, the splendid results of antisepsis in surgery, are due to the protection obtained against destructive Bacteria, there can be at the same time little doubt that this protection, partly due to the absence of germs through the increase of cleanliness consequent on these modes of procedure, is chiefly secured by staying the development of the germs and in a much less degree by their destruction. The elaborate experiments of Koch (14, p. 234) show that of the various disinfecting and antiseptic agents in the proper state of concentration or dilution, only corrosive sublimate, chlorine and bromine have the effect of killing the germs. Bodies like salicylic, carbolic, and other acids in the suitable state of dilution, and powdered cane-sugar can only be supposed to have the desired effect by stopping the growth of the Bacteria. It would be highly important to enquire more closely into the specific sensibilities which may exist in the different species of Bacteria. The behaviour of a Micrococcus of ulcer or erysipelas in the presence of antiseptics may possibly be different from that of Bacillus Anthracis, which has been the chief subject of Koch's study.

VII.

Relation to and effect upon the substratum. Saprophytes and Parasites. Saprophytes as exciting decompositions and fermentations. Characteristic qualities of Forms exciting fermentation.

The vegetative process in organisms, which use organic compounds for their food, must necessarily effect changes in the substratum from which this food is withdrawn. To these changes are added other effects, more closely connected with the process of respiration, which lead to profound transformations in the organic substratum.

This is especially the case with organisms whose mode of life is of the kind described and therefore with all that do not contain chlorophyll, Infusoria and Fungi as well as Bacteria. Fungi, especially in the narrower use of the word, Sprouting Fungi, moulds, &c. being comparatively easy to examine, have supplied the best and most numerous conclusions with respect to the phenomena in question, and we shall often have to make use of them as examples in the following remarks.

The interest attaching to the Bacteria which are devoid of chlorophyll rests chiefly on their effects on their substratum, and after the foregoing introduction we must proceed to consider these organisms, and endeavour to give a clear idea of them by calling attention to the most important known examples.

Organisms not containing chlorophyll are separated into two primary divisions, according as the organic substratum is a living or a dead body. Those which have their habitat on or in living fellow-creatures, and derive their sustenance from them are termed parasites; the others which live on dead bodies are known as saprophytes. Different species are in fact differently adapted to one or the other mode of vegetation; some are known both as parasites and saprophytes, others only in one or the other character.

We shall subsequently have to go more deeply into these distinctions and gradations, especially in the case of parasites. This brief mention of them is sufficient for the present.

The particular account of these forms will be simpler and more intelligible if it begins with saprophytes. The organic compounds present in bodies inhabited by saprophytes are split up into simpler substances; in extreme cases total oxidation, rotting, takes place with the decomposition of non-nitrogenous carbon-compounds into the final products of carbon dioxide and water; in other cases we have partial oxidations, not proceeding so far as the final products of combustion, ' oxidation-fermentations,' as for example in acetous fermentation—that is, the formation of acetic acid by the oxidation of

F

ethyl-alcohol. Reductions are of rarer occurrence, as in the
splitting of sulphates by Beggiatoa, which will be described
presently. The last to be mentioned are the splittings which
end in other than simple products of oxidation and are included
under the general term of fermentations; of these the best-
known example in every respect is the alcoholic fermentation,
which is the splitting of the different sugars into ethyl-alcohol
and carbonic acid. If these splittings are accompanied with a
development of offensive gas, especially in compounds containing
nitrogen, the term putrefaction is used, an expression rather
popular and expressive than strictly and scientifically defined.

It is no part of our subject to enter further into the chemical
nature of these processes, the purely chemical and physical
sides of the theories of fermentation. With regard to the
general history of these theories also, we shall only observe
that it has been an established scientific truth since about the
year 1860, that the entire series of phenomena of rotting
and fermentation above mentioned are the results of processes
of life and vegetation in certain lower organisms, especially
Fungi and Bacteria. To Pasteur belongs the entire credit of
having placed this vitalistic theory of fermentation on a firm
basis, in opposition to other views which acknowledged no
causal relations at all between it and living organisms or causal
relations of a different kind, and of having extended it to all
phenomena of a similar kind. It is true that the same
vitalistic theory has been distinctly expressed in the case of
alcoholic fermentation since the time of Cagniard-Latour (1828)
and Schwann (1837), but it never obtained general accep-
tation.

The vegetative process of living organisms is then the direct
cause of fermentations; there is no fermentation if the organisms
are destroyed. Organisms of this kind are therefore termed
fermentation-exciters, ferment-organisms, or simply ferments
in the terminology of the school of Pasteur. In that of Nägeli
they are known as yeast, and according as the ferment-organism

is a Sprouting Fungus, a Fission-fungus, that is a Bacterium, or a Filamentous Fungus, it is shortly termed Sprouting Yeast, Fission Yeast, or Filamentous Yeast. The French system of terminology limits the application of the French word levûre, which had originally the same meaning as the German Hefe and English yeast, to the Sprouting Fungi which excite fermentation. It is essential to the understanding of the literature to observe that the German Hefe, English yeast, is used in quite different senses ; it must be added also, that the same word is applied not only to the ferment-organism simply, or to the particular form of Sprouting Fungus which excites fermentation, but also to all forms of Sprouting Fungi whether they excite fermentation or not, thus often causing very needless confusion.

We shall speak again presently of the different meanings of the word ferment.

Since the vegetation of organisms sets up fermentation, the substratum in which the fermentation is to take place must contain all the nutrient substances necessary for the process of vegetation. A pure saccharine solution, for example, does not ferment if a small quantity of fermentation-exciting Fungi or Bacteria also in a pure state is introduced into it. The sugar, as we have seen, is a good nutrient material for these organisms. But it only supplies the necessary amount of carbon, the elements of water, and free oxygen, and is therefore imperfect as food. It is only when the compounds which supply the nitrogen mentioned above and the ash-constituents are added to the solution that it is rendered capable of fermentation, and fermentation commences as soon as the conditions favourable to vegetation are secured. Bodies which in the natural course of things or when artificially prepared have finished fermenting, such as must or brewers' mash, are proper food for ferment-organisms.

In every process of fermentation there is first of all a growth, a multiplication of the exciting organism at the expense of the fermenting substance. This can be seen by direct observation

when the smallest possible quantity of the organism is introduced in the beginning, and its weight exactly determined.
The rest of the substratum is split up into the products of
fermentation in consequence of the processes of decomposition
which are connected with the vegetation, and which, as has
been already said, cannot be further considered here. The
best-known example of the kind is the alcoholic fermentation
of sugar by the Sprouting Fungus of beer-yeast, Saccharomyces
Cerevisiae, though it certainly does not strictly belong to the
subject-matter of these lectures. Pasteur states that in a suitable
solution about 1·25 per cent. of sugar was used for the formation
of yeast-substance, 4–5 for that of succinic acid and glycerine,
the remainder, 94–95 per cent., was broken up into alcohol and
carbonic acid.

This example shows that the process of decomposition is
complex, and does not simply consist in the breaking up of all
the sugar into carbonic acid and alcohol. But these, in point
of quantity and from their importance to human requirements,
are the most prominent products of the fermentation in question. Accordingly, we distinguish in this and all other cases
primary and secondary products of fermentations, and we name
the particular process of fermentation from some characteristic
primary product.

It is known that the nature of the fermentations set up by
Bacteria is in general analogous with that of the case just mentioned. But in most of them the splitting-process is at present
less exactly understood, and in many only the primary products
are qualitatively known. Among these carbonic acid constantly
makes its appearance, as in Saccharomyces. Further remarks
will appear below along with special examples. At present we
will only briefly call attention to the colouring matters which
are observed not unfrequently in fermentations with Bacteria;
they were noticed before on page 4, and have given rise to the
expression pigment-fermentations.

Some, but not all, ferment-organisms give off into the fluid medium dissolved substances, which in the very minute quantity in which they are excreted are able to give rise to other changes in the substratum than those which belong directly to the process of fermentation. Analogous products with analogous effects are often obtained from other sources also, for instance in Fungi which do not excite fermentation, and on certain organs or in the cells of higher organisms, even of plants containing chlorophyll. The Fungus of beer-yeast, for example, Saccharomyces Cerevisiae, excretes a substance which inverts cane-sugar in solution, as the phrase is, that is by absorbing water splits it into glucose and laevulose (grape-sugar and fruit-sugar). By means of a similar excretion Bacillus Amylobacter breaks up cellulose into products soluble in water. The cells of germinating seeds produce a body, diastase, which breaks up starch-granules into dextrin and maltose. Substances of this kind are known as enzymes or unformed or unorganised ferments, in German terminology simply ferments. The terminology of the French schools consistently carried out, especially by Duclaux, terms them generally diastases, and then for the separate cases invents special words, all having the same ending, as amylase, saccharase ('sucrase'!), casease, and so on, reserving the word ferment, as we have learned, for the living ferment-organisms themselves. Enzymes, as has been already intimated, are either unorganised bodies or bodies with a definite form, soluble in water, and are all allied as regards chemical character to the proteid compounds. They can with proper management be separated from the organisms which produce them without putting an end to their activity. Their characteristic mark as a rule is the power which they possess of causing chemical changes, chemical separations, without passing themselves into the final products of these changes and so losing their active powers. Their effects are specifically different in every case, and they are accordingly distinguished, as in the examples cited, into inverting, sugar-forming, and other enzymes, to which may be added those that, like the pepsin of the gastric juice of animals, convert albuminous bodies with absorption of water into easily soluble peptones, peptonising enzymes.

After what has now been said it scarcely requires to be pointed out that every organism which sets up fermentation or decomposition displays a specific activity in the directions indicated, and it may be also a specific formation of enzymes. In the same saccharine solution one species excites alcoholic fermentation, another lactic acid or butyric acid fermentation, and so on. Again, the same fermentation, according to the primary products, may also be produced by dissimilar species under otherwise similar conditions, though in unequal quantitative amount. Alcoholic fermentation, for example, is excited in saccharine solutions by several species of Saccharomyces, and also by certain species of the group of Mucorini. The same species can also set up different decompositions in different substrata. The vinegar-bacterium oxidises the alcohol in a dilute solution, and converts it into acetic acid and this into carbonic acid and water when the alcohol is exhausted. The Saccharomyces of beer-yeast changes grape-sugar by fermentation directly into carbonic acid and alcohol; cane-sugar does not ferment, but is first 'inverted' by the above-mentioned enzyme, and the 'invert-sugar' formed of glucose and laevulose ferments as it arises.

The Bacillus of butyl-alcohol of Fitz (Bacillus Amylobacter, see Lecture IX) vegetates in nutrient solutions of milk-sugar, erythrite, ammonium tartrate, salts of lactic acid, malic acid, tartaric acid, &c., without exciting characteristic fermentations in them; it produces fermentation in glycerine, mannite, and cane-sugar, with carbonic acid, butyric acid, and butyl-alcohol as the primary products, and small amounts of lactic and other acids as secondary products, the quantities of the primary products varying much according to the nature of the substratum. The relative quantities of butyric acid, for example, under similar conditions of fermentation, are 17·4 in the case of glycerine, 35·4 in that of mannite, and 42·5 in that of cane-sugar.

Many similar examples are to be found in works on fermentation.

The production of enzymes may also vary in the same form according to the quality of the substratum. Wortmann (34) found in the case of a Bacterium which he does not further determine, that it excretes a starch-dissolving enzyme, and dissolves starch if carbon is presented to it in the form of starch-grains only. If the carbon is offered it in the form of a carbohydrate which is readily soluble in water, such as sugar, or of tartaric acid, the starch-grains which are offered to it at the same time remain untouched. Similar facts are recorded of Bacillus Amylobacter, which, according to van Tieghem, when fed with glucose, leaves the cellulose which is presented to it at the same time untouched, but decomposes it and takes it in as food if no source of more readily assimilable carbon is available.

Lastly, the definite activity of a particular species in the way of fermentation or decomposition may be reduced to zero by a change in the external conditions within the limits of vegetation, even when the quality of the nutrient material remains the same. Examples of this are furnished by the Mucorini already mentioned in passing, by the different species of Saccharomyces, and by Bacillus Amylobacter and other Bacteria. Bacillus Amylobacter, according to Fitz, loses the power of causing fermentation, without losing that of vegetation, when exposed to a high temperature, for instance, after the spores are boiled from 1–3 minutes in a solution of grape-sugar, or after being heated for 7 hours up to 80° C.; the same effect is produced if it is cultivated during many generations with a copious supply of oxygen in a nutrient solution, in which it is unable to excite fermentation. The Mucorini present themselves in very different forms according to the change of conditions, though the form is quite fixed in each particular case. Such a change of form does not occur in Saccharomyces and the Bacteria which have been more thoroughly examined as to this point, or only to an inconsiderable degree. That external conditions of every

kind should have some influence on the form of Bacteria is a
legitimate a priori assumption, and may be directly observed
from the facts stated on page 29.   It is therefore highly
probable, though further distinct proof is required, that the
change of form of strongly pleomorphous Bacteria (see pp. 25–6)
is to a large extent determined by changes in the external
conditions of vegetation.

In the natural course of things the processes of development
and decomposition of which we have been speaking, seldom if
ever go on their way purely and smoothly from beginning to
end.   Many of the organisms in question are so numerous that
their germs find their way simultaneously or in rapid succession
into a nutrient solution or other decomposable substratum.   In
that case they either develope simultaneously and the effects of
their decomposing action appear side by side ; or some find a
favourable substratum at first, but changing its character by their
vegetation, which is thereby impeded, they thus prepare a highly
favourable substratum for other forms ; in this way various de-
velopments and decompositions make their appearance, one
after another, in the same substratum.

Examples of such combinations and successions of products
of fermentation and decomposition are found everywhere in the
natural course of things, and in matters connected with domestic
economy.   There is less need for me to dwell on them here,
because many of them will have to be noticed in the succeeding
descriptions of the several species.

VIII.

**Most important examples of Saprophytes.   The nomen-
clature explained.   Aquatic Saprophytes : Crenothrix,
Cladothrix, Beggiatoa; other aquatic forms.**

In proceeding now to the special consideration of a few
saprophytic Bacteria, three remarks must first be made.   First,

we cannot attempt to give an account of all the phenomena which have been described. We confine ourselves to such as are at present best known, and are at the same time of more general interest. It is to be presumed that many more will have to be added to these in the course of time, and that various changes will have to be made in the views at present entertained. We are still very much in the position of beginners as regards our knowledge of these matters and our investigations. Secondly, we do not propose to go any further into the details of the chemical processes attending the work of decomposition ; we are chiefly concerned with the morphological and biological points of view. Thirdly, we must keep clearly in mind that our knowledge of the morphology and biology of the Bacteria is at present very imperfect, or at least very unequally developed. So much is this the case, that we are not yet in a position to attempt a consistent classification and nomenclature on the principles of systematic botany. What at present seems like such a classification is only a temporary expedient. In such a case the only thing to be done is to agree upon a provisional arrangement and nomenclature for the time being. We will therefore, first of all, adhere to the primary division into endosporous and non-endosporous or arthrosporous forms proposed in Lecture III. Single better-known groups in these two divisions may and must then be constituted genera, and receive names capable of being precisely defined. We limit the use of the name Bacillus, and apply it to all endosporous forms and species with rod-like vegetative cells and cell-unions of the first order. Single arthrosporous forms, such as Beggiatoa, Cladothrix, Leuconostoc, Sarcina and others, may be separated from the rest and distinguished by characters which will be described presently. There still remain a number of forms, in respect of which we are reduced to superficial distinctions of shape, and their ultimate classification must therefore be deferred. Among these the spiral forms may be included under the name Spirillum. Some of these, according to van Tieghem, belong to the endosporous division; others appear to be arthrosporous, while there is a third group in which the point is not yet ascertained, but appearances are in favour of their being kept together, as at present, under the same genus. The rod-forms which are not known to be endosporous may all be termed Bacterium, and the coccus-forms (page 9) Micrococcus. It is obvious that no sharp line of distinction can be drawn between Micrococci and short rod-shaped Bacteria, but it is convenient and customary to distinguish them. The species too, which are at present distinguished, require care in their determination. Some of them are certainly known to be fully and clearly distinct; of others this cannot be said, and their present names in all probability include two or more species which have yet to be studied severally. Thus it seems to me quite certain that more than one distinct species has been described

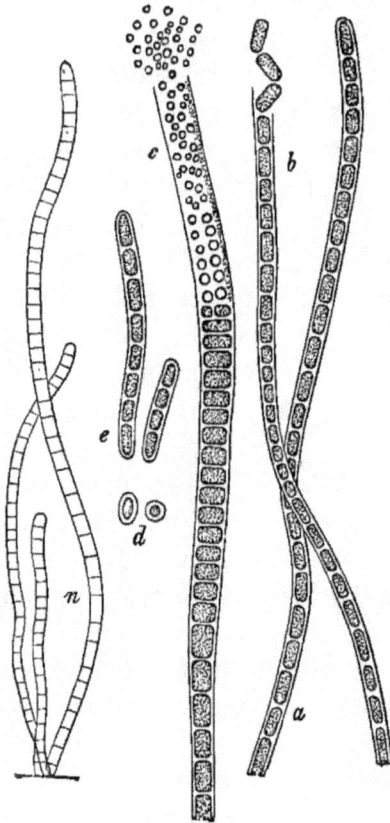

Fig. 5.

Fig. 5. Crenothrix Kühniana, Rabenhorst. *n* group of young filaments attached below. *a, b* older filaments; at the upper end of *b* single cells are issuing from the opened sheath. *c* broad filament with flatly disk-shaped cells in its upper portion, which are divided in basipetal succession along the length of the filament into minute round spore-cells; the spores are issuing from the uppermost extremity of the open sheath. *d, e* spores developing into young filaments. After Zopf. *n* magn. 450, *a, b* 540, *d, e* 600 times.

under the name of Bacillus subtilis. Such collective names—collective species, as we may shortly say—have occurred in all branches of natural history and have been gradually disentangled; here, too, they will ultimately be cleared up. We have only to keep an eye upon them, and not be induced by names to adopt premature conclusions respecting them (35).

We will now proceed to give some examples.

The comparatively large arthrosporous forms, which are described under the names of Crenothrix, Cladothrix, and Beggiatoa, are found often in injurious, or at least in very disagreeable quantities, in waters containing organic substances in solution (36).

1. Crenothrix Kühniana, Rabenhorst (Fig. 5), in the most highly differentiated stage of its development, forms filaments, according to Zopf, $1-6\,\mu$ thick and about 1 cm. long, attached at one end to fixed bodies, entirely unbranched, straight or less often slightly spirally twisted. The filament consists of a row of cylindrical cells, which are half to about one and a half times as long as broad. The outer layers of their lateral walls coalesce and form a delicate sheath surrounding the whole filament, which is colourless when young, but at a later period is often coloured from yellowish to dark brown or brownish green by salts of iron. The filaments not unfrequently break up transversely into pieces, which float free in the water and collect into flocculent masses. The segments of the filaments may pass by repeated bipartitions into the form of isodiametric cells which then round themselves off. In this way the cells of thicker filaments first take the shape of flattish disks, and then divide one or more times in the longitudinal direction of the filament into small roundish cells (*b, c*). These ultimately escape from the sheath, either because the sheath swells up along its whole length, or because it swells up and opens at the apex only and allows the small cells to escape at that point; the cells are either passive and are thrust forth by the continued growth in length of the lower portions of the filament, or have a slow movement

of their own. These minute cells may be called Cocci from their form, or spores on account of their capability of further development, for when cultivated in bog-water they develope into new filaments resembling the parent-filaments (*d*, *e*). On the other hand they may retain the Coccus-form and multiply, producing at the same time a large quantity of jelly, and in this state they form Zoogloeae, which vary in size from microscopical minuteness to more than 1 cm. in diameter. They also occasionally pass, according to Zopf, into the motile condition, and back again into the resting-state. The Zoogloeae are at first without colour, but like the sheaths of the filaments they gradually become coloured by deposition of iron. The Cocci also may ultimately develope from the Zoogloea-state into the filaments as at first described. The external conditions for these formations are not certainly understood.

Crenothrix Kühniana is found in every kind of water, even in the water of the soil as far as twenty metres below the surface. It may become a formidable nuisance in water-pipes, drainpipes, and the like, in which its tufts of filaments and its Zoogloeae increase to such an extent as to form dense gelatinous masses stopping up the passages; in reservoirs it may form slimy layers several feet in depth. The water is thus rendered unfit for drinking and for various technical uses, though no direct injury to human health has been traced to the Crenothrix. We do not know that any other processes of decomposition are caused by Crenothrix.

2. Cladothrix dichotoma, Cohn is of still more frequent occurrence than Crenothrix, especially in dirty water, such as the outflow from manufactories and from similar sources, and also in streams (Fig. 6). It often forms extensive films of flocculent matter of a grayish white colour floating near the edge of the water. Its delicate filaments, ensheathed as in the preceding species, are chiefly distinguished from those of Crenothrix in the full-grown state by being branched. Branching is effected by any single cell of a filament bending one of its extremities

laterally out of the line of the rest, and then growing on in the divergent direction and dividing transversely. The divergent branch forms an acute angle with the primary filament, and in relation to the point of attachment or base of the latter the angle is usually open upwards, seldom the reverse. This form of branching, which is of common occurrence in the Nostocaceae, in Scytonema, for example, and Calothrix, has been termed false branching, because the part which the individual cells take in it, morphologically speaking, is not the same as in most of the other lower plants which have filaments formed of a single row of cells; it is false only in this sense, and is really a peculiar mode of branching.

Whatever else is known of the structure and development of Cladothrix, especially since Zopf's researches, so far agrees with the accounts given of Crenothrix, that only a few remarkable particulars need be touched upon here; Zopf's monograph should be consulted. First of all it is not perhaps superfluous to remark that Cladothrix also receives a deposit of iron oxide in the sheaths of its filaments, and becomes

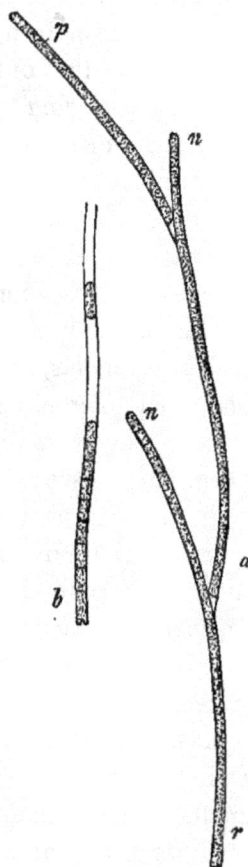

Fig. 6.

Fig. 6. Cladothrix dichotoma, Cohn. *a* extremity of a live filament, which grew originally in the direction *r—p*. The branches *n, n* have been formed by lateral divergence and subsequent growth of segment-cells in the new direction. The construction of the filament out of cylindrical segment-cells is clearly shown at the apex of the branches; elsewhere it is recognisable only by the aid of reagents. *b* portion of a filament showing the segmentation and the sheath; the latter is empty in its upper half, except where one cylindrical cell remains fixed in it. Magn. 600 times, but made a little too broad in the drawing.

coloured accordingly. The often striking accumulations
of ochre-coloured slime-masses in springs and small streams
which contain iron, the filamentous constituents of which are
known by the old name of Leptothrix ochracea, Kützing,
consist, according to Zopf, of this iron-containing Cladothrix.

The filaments multiply by the abscision and further growth
of portions, which form longer or shorter rods according to
their size—a mode also very common among the allied Nostocaceae—and also, according to Zopf, by means of spores or
'Cocci,' that is, short rounded cells, which issue from the sheath
and develope into filaments.

The filaments, or single branches of them, instead of retaining
the usual tolerably straight form, may become spiral with more
or less narrow or open coils, and these spiral forms also may
break up transversely into separate pieces.

Both the longer and the shorter abscised rod-shaped and
spiral portions of filaments, and the round spores and Cocci
also, not unfrequently become motile, the longer ones
creeping or gliding with a slow movement, the short forms
displaying an active swarming motion, such as is described on
page 7.

Lastly, the four forms, the filamentous, the rod-like, the spiral,
and the coccoid, whether mixed together or separate from one
another, may remain united by a jelly into Zoogloeae, which
sometimes appear as bodies of considerable size with shrub-
like branching. The short forms may again become motile,
and swarm out of a Zoogloea; but they may also develope
again into the filamentous form, the typical form from which we
set out; this has certainly not been directly observed in the case
of the spiral rods.

If all these statements are correct, Cladothrix supplies the
most complete example of a pleomorphous course of development.

No more is known of injurious properties and decomposing
power in Cladothrix than in Crenothrix.

3. The species of Beggiatoa (Fig. 7) agree closely, according to Zopf, with Crenothrix and Cladothrix in their pleomorphous course of development. Straight and spiral filaments, abscised straight and spiral rod-like portions of filaments, the latter provided with cilia and described under the name of Ophidomonas (*d*), round Cocci or spores (*e–k*) and Zoogloea-aggregates of these, make their appearance in just the same alternation as in the two preceding genera, rods, Spirilla, and Cocci having in many cases a swarming motion. The distinction between them and the species of Crenothrix and Cladothrix lies chiefly in the presence of sulphur in their structure, and in the motility of the filaments which, like those of Crenothrix, are never branched.

Beggiatoa alba, Vaucher, the most common species, has colourless filaments, attached when quite intact to solid bodies but easily breaking off from them and thus set free, and varying in thickness from 1 to 5 μ. The filaments consist of cells of more or less elongate cylindrical to flat disk-like form, the latter occurring especially in the thicker specimens. They have no distinct sheath clothing the row of cells; moreover, while the protoplasm of Crenothrix and Cladothrix is uniformly clouded or finely granular, in Beggiatoa alba it has disseminated through its substance comparatively thick round highly refringent grains, with a dark contour therefore, and composed of sulphur, as Cramer has shown. Similar sulphur-grains are also present in the non-filamentous states or forms assigned here by Zopf. Their number is not the same in different filaments; in some filaments (*c*) but few are to be seen, and in parts of them they may even be entirely wanting. In most filaments they are present in large numbers, so large sometimes that they entirely conceal the structure of the thread, which looks like a rod having its uniformly clouded protoplasm traversed by a dense mass of granules with a black outline. It is only by the use of reagents which largely withdraw the water of the cells that it is possible to distinguish them (*b*).

Again, the filaments usually exhibit active movements, such as are known in the green Oscillatorieae, which have been noticed already several times, and which are undoubtedly the near allies containing chlorophyll of the species of Beggiatoa and of the arthrosporous Bacteria. The movements consist in progression in the line of the axis of the filament in one direction, or in opposite directions alternately, together with rotation in a path which forms the outline of a very pointed cone, or of a double cone such as is described in the case of rod-shaped Bacteria. These movements, when hastily observed, appear to be gliding in a forward direction, while the ends of the filaments swing hither and thither in the manner of a pendulum. Sometimes also the filaments become curved, and then often straighten themselves again with a jerk, showing their great flexibility throughout their entire length.

Several other species of Beggiatoa are known: B. roseo-persicina, distinguished by its rose-red to violet colour, and also said to be pleomorphous, its Zoogloeae, according to Zopf, being Cohn's Clathrocystis roseo-persicina; B. mirabilis, Cohn, known only in the filamentous form, a gigantic species 20–30 μ in thickness; B. arachnoidea, Roth, and some others. Apart from the differences indicated, all these agree with B. alba in the characteristic marks, especially in the presence of sulphur-grains.

B. alba is one of the most common inhabitants of our waters. It is found in the water of marshes, in the waters that flow from manufactories, in hot sulphur-springs, and in these places often in company with Cladothrix, and in the sea on shallow coasts. B. roseo-persicina is less common in these localities; the other species mentioned above are known only as coming from the sea. The species of Beggiatoa live on the decomposing remains of organised bodies, especially plants; they are, therefore, chiefly found at the bottom of water, where such objects accumulate. They form there, when largely developed, slimy membranous coverings or films of flocculent matter, which are either white in colour or vary from rose to brown-violet, as in B. roseo-persicina.

The species of Beggiatoa are said to have the peculiar power of reducing the sulphates contained in the waters which they inhabit, especially sodium sulphate and gypsum, setting free the sulphur and sulphuretted hydrogen.   That the living protoplasm is the seat of this process is shown by the appearance in it of the sulphur-grains.   The form-ation of sulphuretted hydrogen causes first the precipitation of iron sulphide in the slime inhabited by Beggiatoa, which is thereby turned black, and then the presence of sulphuretted hydrogen, either dissolved in the water or set free by evaporation, gives rise to the well-known odour, and may have a noxious effect on the animals inhabiting the water. The ' white ' ground in the Bay of Kiel, for example, covered by species of Beggiatoa, is also called the ' dead ' ground, because it is avoided by fishes, though not by all animals (37). These plants therefore play a peculiar and important part in the economy of nature and of mankind. According to the statements of some

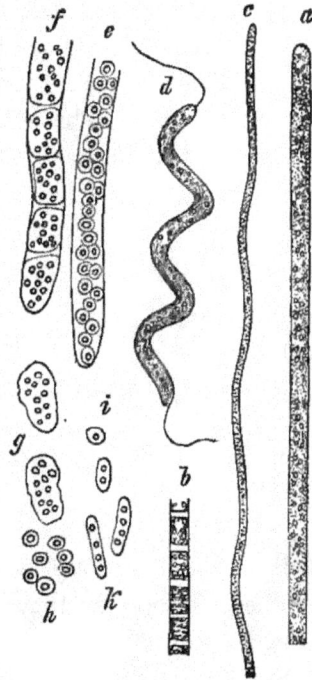

Fig. 7.

Fig. 7. Beggiatoa alba, Vaucher. *a* portion of a stout living filament. *b* fragment of the same after treatment with alcoholic solution of iodine show-ing the segmentation into cells. *c* a very thin living filament from the same preparation as *a*. *d* motile spiral form (Ophidomonas). *e–h* formation of spores ('Cocci') by successive division of the segment-cells of a stout filament (*e*). The lumen of each spore is nearly filled up by a grain of sulphur. In *f* the division has advanced further than in *e*. *g* breaking up of the filament into groups of spores. *h* the spores isolated. *i, k* spores appearing to germinate ( ), in a state of motion. *a–c* magn. 600 times, but drawn a little too large, *d* 540 times, *e–k* 900 times. *d–k* after Zopf.

G

observers, they share this part with other plants which are green
and are related to the Oscillatorieae and Ulothricheae (25, p. 769).

The forms which have now been described are the largest,
but by no means the only, representatives of the aquatic Bac-
terium-flora.

The Spirilla which live in bog-water are remarkable forms,
and may be briefly illustrated here by two examples. Spirillum
Undula, Cohn (Fig. 8, *A*), forms small spirally-twisted rods of about
1 μ in thickness. The width of the spiral in dead specimens is
about 3 μ, three times therefore the diameter of the cell, the height
of a turn of the spiral 5–6 μ
(4–5 μ according to Cohn).
Each individual is usually
formed of from $1\frac{1}{2}$ to 2
turns only of the spiral;
when it has reached this
length it divides transverse-
ly in the middle into two. According to Cohn 3 turns of the
spiral are only rarely attained. The rod consists of segment-
cells, which, as far as can be determined, are immediately after
division about as long as a half turn; they separate from one
another as soon as they are of this size (*a*), or after longer
growth.

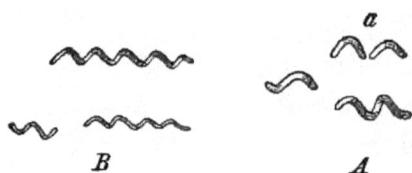

Spirillum tenue, Cohn (Fig. 8, *B*), is more slender and more
closely twisted than S. Undula, and has several connected turns
of the spiral, usually 3, 4, or 6. The length of each of the
segment-cells which compose the spiral is at the time of division,
as far as I was able to determine, about half a turn, the same
therefore as in S. Undula.

No other phenomena of development than growth and division
of the rods have been perceived in these two species, even during
a cultivation of some months' duration; they both remain

Fig. 8.   *A* Spirillum Undula, Cohn; at *a* separation into two segment-
cells. *B* Spirillum tenue, Cohn; three specimens of different lengths.
Magn. 600–700 times.

constant in their forms and distinctions. They are often found by themselves in the waters of bogs; when they occur in large quantities, and comparatively unmixed with other species, they form dense swarms which, in S. Undula especially, are of a beautiful dark reddish-brown colour. Single live rods are, under the microscope, colourless and homogeneous. When killed and treated with colouring reagents (iodine, anilin-dyes), they exhibit a remarkable separation in the case of both species into short irregular transverse zones of alternately darker and lighter colour—a phenomenon which must not be confounded with the transverse segmentation into distinct cells mentioned above. Finally, both species are distinguished by the extreme vivacity of their movements, and dart like meteors, says Cohn, across the field of vision, the slender Spirillum tenue affording in this way a very elegant display.

Many other forms of the kinds previously described, and among them of endosporous Bacilli, might be mentioned as living in water. We still desiderate such investigation of these forms as would enable us to give a more exact account of them and of the decompositions which they may effect; the scattered particulars which are known of them have no interest for us on the present occasion. Of the germs of Bacteria which may be found even in the purest waters when exposed to the air and to dust I have already spoken in the fifth Lecture.

IX.

Saprophytes which excite fermentation. Fermentations of urea. Nitrification. Acetous fermentation. Viscous fermentations. Formation of lactic acid. Kefir. Bacillus Amylobacter. Decompositions of proteid. Bacterium Termo.

WE have now to consider saprophytic forms which are known to be the causes of distinct processes of decomposition or fermentation, and as examples of more general interest we select for closer examination the Micrococcus of urea, the nitrifying Bacteria, the mother of vinegar, the Bacteria of lactic acid-fermentation, of butyric acid-fermentation, and of the viscous fermentation of carbohydrates and other bodies, and, lastly, the Bacteria of the decompositions of proteids.

1. The normal urine of men and carnivorous animals if kept exposed to the air acquires an alkaline and ammoniacal smell in place of the acid reaction present in it when fresh. The cause of this is, that the urea takes in water and is converted into ammonium carbonate. The originally clear fluid becomes clouded by the presence, as examination shows, of a number of lower organisms, among which there may be a variety of Fungi and Bacteria. Pasteur first proved that one of these Bacteria, Micrococcus Ureae, Cohn, was the exciting cause of this process of fermentation in urea (38; 25, p. 697), by showing that the Micrococcus, if grown pure and cultivated in a pure nutrient solution containing urea, causes the same decomposition in it as in urine.

The Micrococcus (Fig. 9) consists of small round cells about 0.8 μ in diameter, which usually, though not invariably, remain connected together in rows often of more than 12 cells. These rows are in many cases curved and bent in an undulating manner, and are often ultimately wound into coils or small Zoogloeae as they may be termed, in which the cells appear to be irregularly heaped together. At the first beginning of a culture the cells, according to von Jacksch, are cylindrical, though not very much longer than they are broad; they retain this shape for some time, firmly united together in genetic connection, forming therefore rod-like rows of short cylindrical cells, and afterwards become rounded off. We may therefore, if we please, speak of a 'rod-form,' but we shall not gain in clearness by so doing. No distinct spores have been observed in this

Fig. 9.

Fig. 9. Micrococcus Ureae, Cohn, from decomposing urine. Cells separate or united in rows (= Streptococcus). Magn. 1100 times.

Micrococcus. Leube has recently demonstrated the existence of four quite distinct species of Bacteria, in addition to the Micrococcus just described, producing the same effects.

Micrococcus Ureae, as we learn from experiment, requires a supply of oxygen for its vegetation. It can hardly therefore be the cause of the alkalisation of the urine inside the bladder, which has been observed in some affections of the bladder and is supposed to be an effect of it, for the oxygen required is not present. But numbers of small Bacteria are found in urine in this diseased alkaline condition, and it must be assumed that having found their way spontaneously or forcibly, as for instance by means of the catheter, through the urethra into the bladder they are the exciting cause of the decomposition in question. It must accordingly be further assumed, that other species which are anaerobiotic have the power of producing fermentation in urine, or processes similar to it. Leube's species appear from the accounts given of them not to be anaerobiotic. Miquel (15, vol. for 1882) has in fact discovered a very delicate rod-form occurring in dust, which he names Bacillus Ureae and which vegetates anaerobiotically, converting urea into ammonium carbonate in the same way as Micrococcus Ureae.

We learn from van Tieghem that hippuric acid is converted into benzoic acid and glycocol in the urine of herbivorous animals by a Micrococcus, which is perhaps identical with Micrococcus Ureae, but requires further investigation.

2. In connection with the forms which change urine into ammoniacal compounds, we may now turn to the consideration of nitrification, the oxidation of compounds of ammonium into nitrates, such as occurs on a large scale in the formation of saltpetre, in so far as this also is due, according to the observations of Schlössing and Müntz (25, p. 708; 39), to the vegetation of small Bacteria. The phenomenon occurs in moist soil penetrated by air and containing compounds of ammonium with small quantities of organic matter and basic substances, for example, salts of calcium. It may be induced artificially in

nutrient solutions containing compounds of ammonium, if a small quantity of soil is added at a suitable temperature, the optimum being 37° C., and with constant access of air.

Thus the formation of saltpetre is a result of the vegetation of Bacteria; it ceases when these are killed; it also commences when these Bacteria, artificially reared, are placed by themselves without soil in the proper nutrient solution. From this we must conclude that we have here an oxidation produced by the Bacteria which are widely diffused in the superficial layers of a moist soil.

The morphology of these Bacteria is not yet clearly ascertained. According to the above-named writers, the individual organism is a very small delicate Micrococcus, somewhat resembling M. aceti, and van Tieghem, in his text-book, has named it M. nitrificans. But the appearance assumed by this form is not clear from the descriptions, and Duclaux speaks of a mixture of different forms. The importance of the processes calls for a more exact study of them, that is, of the question, whether nitrification is the exclusive function of a distinct species, or of several species and their combinations.

3. Acetous fermentation (25, p. 504; 40, 41, 42). If an acid nutrient solution containing a small percentage of alcohol is exposed to the air at a temperature of about 30–40° C., vinegar is formed in it, that is, the alcohol is oxidised into acetic acid. The fluid is at the same time more or less clouded, and its surface covered with a thin colourless membrane. This membrane consists in most pure cases of mother of vinegar, Micrococcus aceti, Bacterium aceti (Arthrobacterium aceti; Mycoderma aceti in Pasteur's earlier nomenclature). Pasteur showed twenty-five years ago that this Bacterium lives and grows on the organic and mineral substances contained in the solution, and absorbing oxygen from the air oxidises the alcohol into acetic acid. The exact proof was obtained by adding 4 per cent. of alcohol and 1–2 per cent. of acetic acid to pure nutrient solutions of the kind described on pages 61 and 67, and then introducing into the liquid an infinitesimal quantity of membrane of mother of

vinegar. In the proper temperature, and with free access of
atmospheric air, the mother of vinegar developes into the membrane described above, and as this takes place the alcohol in
solution is converted into acetic acid.

The various methods used in the arts for the preparation of
vinegar, into the details of which we do not here enter, are cultures of Micrococcus aceti at the proper temperature, and with
exposure to the air under regulations which vary in each particular case. The mixtures from which the vinegar is to be
prepared—beer, wine, &c., with addition of previously formed
vinegar, have the essential characters of nutrient solutions as
described above. The vinegar of commerce is a diluted solution
of acetic acid, and contains a larger or smaller number of the
Micrococcus aceti. Germs of this organism are also diffused
elsewhere, and in particular are never wanting in the vessels
used for the preparation and storing of alcoholic fluids. When
these turn acid, owing to careless management, it is in part at
least owing to the activity of the Micrococcus. M. aceti, like
M. Ureae, is as far as we know at present an arthrosporous
Bacterium, and resembles the latter in shape (Fig. 10). It consists usually, and always in the normal vegetating stage, of
cylindrical cells, which are not much longer than broad, and
have a transverse diameter of about 0·8–1 μ. The cells multiply
by the usual process of transverse division, and often remain
united together in rows forming long filaments; in older cultures
they are often thrust out of the filament but are held together
by jelly. With this short-celled Micrococcus-form cell-rows
often occur, in which some cells are in the form of long rods,
others not only several times longer than broad, but also fusiform
and so swollen in a bladder-like manner that their greatest
breadth may be more than four times the diameter of the ordinary
cells. No one would suppose these inflated cells to belong to
the small ones, if they did not usually occur with them, either
singly or several together, as members of the same genetic rows
and connected with them by a variety of intermediate forms.

Cases of this kind have been observed also in other Bacteria; these are the cells with which we made acquaintance before on page 10 under Nägeli's name of involution-forms. Whether they are really retrograde states, as this name would express, or diseased forms, I shall not undertake to say in the case of the Micrococcus aceti. They certainly do not appear at all in some cultures, or only one by one, while in others they are extraordinarily numerous, and in the latter case I could never find that they gave the 'impression of being incapable of further development.' Positive statements, however, are at present no more possible respecting their significance in the history of development than they are respecting the conditions of their presence or absence.

A Micrococcus has been found by E. Chr. Hansen, and named by him M. Pasteurianus, which behaves in every respect in the same way as M. aceti, except that its cells throughout the successive generations show the blue reaction of starch with iodine (see page 5), while the ordinary M. aceti is coloured yellow by that reagent. This fact shows at once that M. aceti although certainly the usual is not the only vinegar-forming species. In fact the power

Fig. 10.

of producing acetic acid has been observed in some other Bacteria, which are, however, comparatively unimportant to us for our present purpose.

Micrococcus aceti has the power not only of producing but also of destroying vinegar. When it has oxidised all the alcohol of a fluid into acetic acid, it may continue to develope, as Pasteur showed, and by a further process of oxidation convert the acetic acid into carbonic acid and water, the final products of all decomposition.

Fig. 10. Micrococcus aceti (mother of vinegar); roundish cells, single and united in rows, also rows of elongated rod-like and fusiform or swollen flask-shaped members; the latter from a culture at a temperature of 40° C. Magn. 600 times.

It does not strictly belong to our subject, but it is perhaps not superfluous to remark that every white membrane which makes its appearance spontaneously on the surface of a fluid suitable for forming vinegar is not necessarily mother of vinegar. The white and ultimately wrinkled film which usually forms on beer or wine, is a well-known object, and to the naked eye looks so like the membrane of vinegar as to be often mistaken for it. But under the microscope it is distinguished from it by being formed from a comparatively large Sprouting Fungus, Saccharomyces Mycoderma, which has no direct connection with the formation of vinegar. On the contrary, it converts alcohol and other bodies in solution by oxidation into carbonic acid and water. Indirectly it may, indeed, in this way promote the formation of vinegar by destroying any excessive amount of alcohol and acid which would impede the development of the Micrococcus aceti, and so providing it with a substratum favourable to its vegetation.

4. We now come to a series of examples of phenomena of fermentation and decomposition produced by Bacteria in the sugars and in the allied carbohydrates. When in the following remarks we speak simply of saccharine solutions, it is always to be understood that they contain also the constituents required for nutrient solutions.

We must first of all say a word or two respecting the so-called viscous fermentations (25, p. 572, 43, 44). The juices of plants which contain sugar, such as onion and beet, when extracted by crushing often assume a sticky viscous character, and produce carbonic acid and in many cases also mannite. Organisms also, to be described presently, make their appearance as a sediment in the viscous mass. If a small portion of the substance is introduced into a suitable solution of cane-sugar which was before free from germs, the same viscidity is caused in it as the organisms develope. These must therefore be regarded as the causes of the change. The organisms in question are, according to Pasteur, of two kinds. The first is a Micrococcus

very like M. Ureae and forming rows of bead-like cells ; by itself
it produces viscidity and mannite in the cane-sugar solution with
separation of carbonic acid.   Secondly, cells of irregular shape
and somewhat larger size than those of the Saccharomyces of
beer-yeast (see page 98), and with morphological peculiarities,
which the descriptions which we have of them do not at all
clearly explain ; these cells are at all events not Bacteria, and
are said to cause viscidity only and to form no mannite in the
cane-sugar solution.   The viscous substance itself, of which we
are here speaking, is stated to be a carbohydrate with the for-
mula of cellulose ($C_6 H_{10} O_5$).

From these data, which it is true are still very imperfect, it
must be acknowledged that the disengaged carbonic acid and the
mannite are products of fermentation ; but the viscous substance
itself is more probably to be placed in the category of mucilagino-gelatinous cell-membranes, which are so common in Bacteria and Fungi and which we have already observed so often
in connection with the Zoogloeae ; it is therefore not a product
of the fermentation of the nutrient solution, but of the assimilation of the organism which excites the fermentation.

This view is distinctly supported by the history of the
development and vegetation of Leuconostoc mesenterioides, the
frog-spawn-bacterium of sugar-manufactories, examined by Cienkowski and van Tieghem, which has the power of converting
large casks of the juice of the sugar-beet in a short space of
time into a mucilagino-gelatinous mass and thus of causing
considerable loss.   Durin saw a wooden vat containing fifty
hectolitres of a 10 per cent. solution of molasses become filled
with a compact Leuconostoc-jelly in less than twelve hours.   The
development of Leuconostoc was mentioned above on page 22 as
an example of an arthrosporous course of development, but a
more detailed description of it must now be given. See Fig. 11.

The round spore-cell (*d*) germinates in a nutrient solution,
and appears at first to be surrounded by a gelatinous envelope
several times thicker than the spore itself (*e*).   Then a simple

filiform row of isodiametric cells is formed by the growth and successive transverse division of the protoplasmic body, and the envelope follows the longitudinal growth of the cells, forming a thick, rounded, cylindrical sheath of the consistence of firm gelatine round the filament. The transverse walls also of the filament in its young state are gelatinous, appearing as broad pellucid partitions between the protoplasmic bodies and being continued into the sheath on the outside (*f-i*). The partitions disappear in older filaments and the protoplasmic bodies are in contact

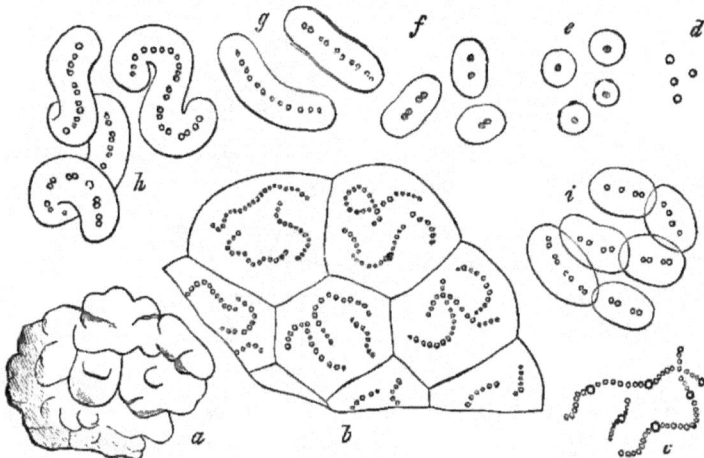

Fig. 11.

with one another (*b*). As the single filament developed from a spore increases in length it forms successively stronger curvatures, which lay themselves in loops round each other and round

Fig. 11. Leuconostoc mesenterioides, Cienkowski. *a* sketch of a Zoogloea. *b* section through a full-grown Zoogloea before the commencement of spore-formation. *c* filament with spores from an older specimen. *d* isolated ripe spores. *e-i* successive products of germination of the spores sown in a nutrient solution. Order of development according to the letters. In *e* the two lower specimens show fragments of the ruptured spore-membrane on the outer surface of the gelatinous envelope indicated by dark strokes. *i* portion of the gelatinous body from *h* divided into short members, and with the members separated from one another by pressure. *a* natural size, other figures magn. 520 times. After van Tieghem in Ann. d. Sc. nat. sér. 6, VII.

other filaments. Growth is also accompanied with separation of the originally elongated gelatinous filaments into shorter transverse sections, which are always surrounded by the sheath and remain firmly attached to one another (*i*). Closely twisted coils are thus produced of the size of or larger than a hazel-nut (*a*), forming the compact gelatinous bodies mentioned above which accumulate and fill the casks. Sections through older gelatinous bodies appear to be divided from the edges of the sheaths into chambers in which the curved cell-rows lie (*b*). When the development is completed, and the nutrient solution exhausted, the gelatinous sheaths deliquesce, the cell-rows separate, and most of the cells die. But previously to this single cells, not occurring in any particular order in the row, develope into distinct spores, becoming a little larger than the rest, and surrounding themselves with a firm non-gelatinous membrane, the outer coat of the spore (*c*). It was from these spores that we set out in the description; but every living portion of a filament that from any cause becomes separated from the connection may develope into a new gelatinous body. The vegetating protoplasmic bodies are, according to van Tieghem, $0.8–1.2 \mu$ in thickness, the sheaths $6–20 \mu$, the spores $1.8–2 \mu$.

In the germination of the spore the gelatinous sheath originates (*e*) as a newly formed inner layer of the cell-wall, or by the considerable increase in thickness of a pre-existing inner layer; the outer coat of the spore then bursts into pieces. This shows decisively that the sheath is a product of assimilation, a growing part of the growing filament. The gelatinous substance has the same chemical composition as the mucilage of viscous fermentation. The material for its formation is of course supplied by the sugar of the solution. In van Tieghem's cultures of Leuconostoc in a solution of glucose, air being admitted and the fluid prevented from becoming strongly acid, about 40 per cent. of the sugar which disappeared was expended in the formation of the Leuconostoc itself; the greater part of the remainder was converted by combustion into carbonic acid

and water without any sensible development of gas. The cultivation of Leuconostoc in solution of cane-sugar soon resulted in
the splitting (inversion) of the sugar into glucose and laevulose,
and for this reason it is so highly detrimental to the fabrication
of cane-sugar; the sugar then disappears as in the first experiment, the glucose first, and about 40–45 per cent. of the sugar
which disappears is expended in the formation of the Leuconostoc.

A similar formation of mucilage to that of the viscous
fermentation of saccharine solutions, is seen in the ropiness
of beer and wine, in which condition they are capable of
being drawn out into filaments. These phenomena also are
accompanied, or doubtless caused, by the formation of Micrococci united together in rows, and the slime may very well have
the same origin and morphological significance as the jelly of
Leuconostoc. It may be observed in passing that other so-
called ailments of beer and wines are caused by Bacteria, but
we cannot enter into any further description of them here[1].

5. The old method of inducing ordinary lactic acid fermentation (25, 45) of the different kinds of sugar is by adding sour
milk or cheese to a fermentable solution and keeping it exposed
to the air at a temperature of 40–50° C. Calcium carbonate
or zinc-white must also be added in order to throw down the
lactic acid as it is disengaged, because the fermentation ceases
as soon as the acid content of the fluid exceeds a certain
amount.

Pasteur first showed that a particular Bacterium, and others
perhaps along with it, was introduced with the cheese or sour
milk, and that it vegetates in the fluid, especially in the sediment
at the bottom, and acts as a ferment. It appears in the form
of minute cylindrical cells, which immediately after division are
scarcely half as long again as they are broad, and average $0.5\,\mu$
in thickness. After each division they usually soon separate
from one another, rarely remaining united, and forming short

[1] See Pasteur, Études sur le vin, Paris, 1866, and Études sur la bière,
Paris, 1872.

rows ; portions of the transverse partition-walls are plainly seen
and are indicated by a slight constriction. They have no power
of independent movement. In form, therefore, this species resembles the Bacterium of vinegar and may be called Micrococcus
lacticus, as has been done by van Tieghem. Hueppe, however,
states that there is a formation of spores, if I understand him
aright, after the endosporous type. If this is confirmed, the
Bacterium of lactic acid is a very small Bacillus in our use of
the term, and must be so named.

It is evident that this Micrococcus or Bacillus is always
present in milk, not of course when it comes from the udder,
but as soon as it is in use. The germs of the organism are
diffused to such an extent in the cattle-stalls and in the vessels
of the dairy that it never fails to be developed. This is why
milk turns sour, because the Bacterium excites lactic acid-
fermentation of the sugar contained in the milk ; and when
the acidification has reached a certain point, the lactic acid
causes the homogeneous gelatinous coagulation of the casein,
which is characteristic of good buttermilk.

Further physiological peculiarities of this Bacterium have been
described in detail in Hueppe's careful treatise, and should be
studied there.

In the Bacillus or Micrococcus lacticus which has now been
described we have made acquaintance with a very widely diffused
and active lactic acid-ferment, but it is by no means the only
one. On the contrary, the number of species of Bacteria which
form lactic acid in saccharine solutions or in milk appears to
be more than usually large. Hueppe alone mentions five, and
all Micrococci ; one of these is known to us as M. prodigiosus of
the blood-portent mentioned on page 14. Two others Hueppe
discovered to be the exciting cause of the lactic acid which
occurs in the human mouth, while he found the Bacillus first
described only occasionally in the mouth. We still wait for
closer investigation into most of these forms, and into their
operation as ferments. It is, however, already plain that on all,

or almost all, occasions on which lactic acid makes its appearance in considerable quantities we may expect to find a ferment-
organism, and indeed a Bacterium which produces it, but this
need not always be the one form, described above as that of
ordinary acidification of milk. It is well to call particular
attention to this point on account of the wide diffusion of lactic
acid, for example in human food, whether this be purposely
made sour, as in ' sauerkraut ' and the like, or is a case in which
the turning sour indicates decomposition, as in soured vegetables
or beer, so far as the effect in the latter case is not due to the
presence of acetic acid.

6. This will be the best place to recur briefly to the Bacterium of kefir mentioned above on page 13, which is connected
with an interesting change in milk. It was discovered by E. Kern
in 1882 (46). Kefir or kephir is the name of a drink, a fluid
effervescing kind of sour milk containing a certain amount of
alcohol, which the inhabitants of the upper Caucasus prepare
from the milk of cows, goats, or sheep, and therefore not to be
confounded with the koumiss obtained by the Nomads of the
Steppe originally from mares' milk, with which we are not at
present concerned. The drink is prepared by adding to the
milk the bodies described above as a beautiful example of
Zoogloeae, which bear the name of kefir-grains. The Cauca-
sians make use in this process of leathern bottles to hold the
milk, the more polished European employs less objectionable
glass vessels. The recipe followed by the latter is mainly as follows.

Living and thoroughly moistened kefir-grains are added to
fresh milk in the proportion of 1 volume of the grains to about
6–7 volumes of milk. The mixture is exposed to the air for
twenty-four hours at the ordinary temperature of a room, protected from dust by a loose covering only, and is frequently
shaken. At the end of twenty-four hours the milk is poured
off from the grains, which may be employed again for a fresh
preparation. The milk itself, which we will term ferment-milk,
is then mixed with twice the quantity of fresh milk, put into

bottles well corked and frequently shaken. The bottled sour milk, which is more or less highly effervescent, is fit to drink in one or more days. It has the somewhat acid taste indicated by its name, and contains an amount of carbonic acid varying according to the temperature and the duration of the fermentation, but sometimes sufficient to burst the bottles or drive out the corks, and, as has been already said, a certain amount of alcohol, which in the cases examined in Germany was less than 1 per cent. but according to other accounts may be 1–2 per cent.

The changes in the milk which produce the drink here described are brought about by the combined activity of at least three ferment-organisms. The kefir-grains, as has been already stated (page 13), consist chiefly of the gelatinous filamentous Bacterium which has been named by Kern Dispora caucasica; intermixed with this organism and enclosed in the tough Zoogloea are numerous groups of a Sprouting Fungus, a Saccharomyces, resembling the yeast-plant of beer; thirdly, there is the ordinary Bacterium of lactic acid, which partly adheres to the grains in company with some unimportant Fungi and other impurities, and partly is introduced each time with the fresh milk.

We know at least enough of the ferment-effects of these organisms or of their near allies to enable us to form a probable idea of the course of the changes which have been described. The acidification is caused by the conversion of a portion of the milk-sugar into lactic acid by the Bacterium of that acid. The alcoholic fermentation, that is, the formation of alcohol and of a large part at least of the carbonic acid, is indebted for its material to another portion of the milk-sugar, and for its existence to the fermenting power of the Sprouting Fungus. The kefir-grain, like its constituent the Sprouting Fungus working by itself, gives rise to alcoholic fermentation in a nutrient solution of grape-sugar, though of a less active kind than that caused by the Sprouting Fungus of beer-yeast. But alcoholic fermentation is produced in milk-sugar as such neither by Sprouting

Fungi with which we are acquainted, nor, as experiment has shown, by those of which we are speaking. To make this fermentation possible, the sugar must first be inverted, split into fermentable kinds of sugar. According to Nägeli (9, p. 12), the formation of an enzyme which inverts milk-sugar is a general phenomenon in Bacteria, and Hueppe has shown that it is probable in the case of his Bacillus of lactic acid in particular ; the inversion required in this case to enable the Sprouting Fungus to set up alcoholic fermentation is the work therefore of the Bacillus of lactic acid, or of the Bacterium of the Zoogloea, or of both.

Lastly, it is to be observed that the kefir is in the fluid state. Coagulation of the casein does indeed take place, but either it is from the first not in the homogeneous gelatinous form of ordinary sour milk but in small lumps and flakes suspended in serum, or the gelatinous coagulations which are sometimes present at first are soon partially dissolved. It is evident then that the casein, which has already been coagulated, is partially liquefied (peptonised). This must be ascribed to the enzyme formed by the Bacterium of the Zoogloea, because according to our present knowledge the Bacterium of lactic acid has no power to peptonise casein or otherwise liquefy it.

This view, which corresponds in all important points with Hueppe's brief communication on the subject, is in accordance with the remarkable fact that the ferment-milk, by means of which the kefir is prepared, contains a large number of actively growing cells of a Sprouting Fungus and of Bacteria of lactic acid, but no Bacteria of the Zoogloea, or only small and doubtful quantities of them. The grains as a rule strictly retain these, while they part with the sprouting cells to the milk. There is obviously no objection to the supposition that enzymes produced from the grains pass over into the ferment-milk and co-operate with it.

In this way, as I have said, we may explain the formation of kefir, and I gave this account of it in the first edition of

H

this work while calling attention to the want of precise investigation.

But A. Levy, of Hagenau, has recently discovered that the effervescing alcoholic kefir may be obtained without any kefir-grains, but simply by shaking the milk with sufficient violence while it is turning sour. A trial convinced me of the correctness of this statement. The kefir obtained by shaking was not perceptibly different in taste or other qualities from the kefir of the grains, and the determination of the alcohol, kindly made for me by Professor Schmiedeberg, gave 1 per cent. in some specimens of the former kind and 0·4 per cent. in one of the latter; sour milk not shaken contained no trace of alcohol or only a doubtful one. Our former explanation therefore must be abandoned, and there is no other ready at present to take its place; but the case is full of instruction for our warning.

Turning now for a moment to the life-history of the kefir-grains, we may briefly remark with respect to the Saccharomyces, that it grows in the sprout-form observed in the Saccharomyces of beer-yeast, partly forming groups or nests inside or on the surface of the grains, partly separating from them and entering the surrounding fluid. It is on an average smaller and narrower than Saccharomyces Cerevisiae, but some idea of its form may be gathered from a figure here reproduced of that Fungus which it very closely resembles (Fig. 12). Of the Bacterium, of which the grains chiefly consist, I believe that we also know only the vegetative development.

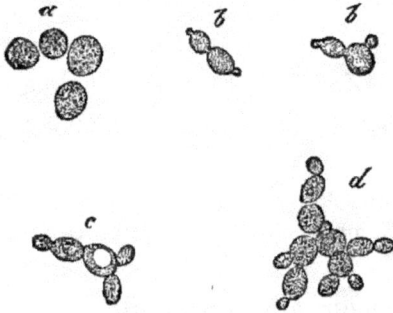

Fig. 12.

Fig. 12. Saccharomyces Cerevisiae. *a* cells before sprouting. *b–d* sproutings in fermenting saccharine solution (sequence of development according to the letters). Magn. 390 times.

It appears, as has been already said, in the form of small slender rods united into filaments, which are closely interwoven and held together in Zoogloeae by means of a jelly.

The source of the grains has not been traced further back than the leather milk-bottles of the mountaineers; their place of birth is still unknown. They come to us in the dry state, and are kept in this manner in the Caucasus also. They must be dried quickly, and the best plan is to dry them in the sun. Much of the dry imported material is dead when it comes, as far as my experience goes. The softened living grain grows slowly in the milk, as we have already seen (page 59), with uniform increase in size and multiplication of all its parts. This growth is accompanied by the separation from time to time of single lobes of different sizes from the whole, and thus the number of the grains increases. From isolated observations I regard it as possible that Dispora-cells sometimes issue from a grain, and may then develope into kefir-grains, but this is not certain. Distinct formation of spores has not yet been observed. Kern it is true has not only described such a formation, but named the Bacterium of kefir Dispora, because two spores are formed each time in a rod, one at each end. After repeated observation I have never seen anything of the kind, though I have very often seen figures which answer to Kern's represent-ations, and which are due to the circumstance that a rod or portion of a filament is curved and its middle portion lying horizontally is seen in its length, but one or both of its extremities which bend away from the horizontal plane are viewed in cross-profile. It is by such appearances that Kern has allowed himself to be misled. If we allow the name Dispora to be used provisionally, it must not be forgotten that the character which it is intended to express does not really exist.

7. We will close the series of examples of the Bacteria which excite characteristic fermentations in non-nitrogenous com-pounds by the consideration of a species of Bacterium which is

one of the most widely diffused, and most important and varied
in its powers of decomposition, the Bacillus of butyric acid,
known as B. Amylobacter, van Tieghem, B. butyricus, Clostri-
dium butyricum, Prazmowski (22, 47, 48), and by some other
names. I think that I ought also to refer Bacillus butylicus of
Fitz to this species, though it must be remembered that this
species as at present established may perhaps be divided into
several on further investigation.

Bacillus Amylobacter (Fig. 13) is nearly 1 μ in thickness, and
vegetates in the form of slender cylindrical rods united at
most into short rows, and usually in a state of active movement.
It is easy to characterise morphologically, because the sporogenous cells swell out till each becomes
fusiform and then produce inside the part
which is most enlarged an elongate ovoid
spore with rounded ends and sometimes
slightly bent, surrounded with a broad
gelatinous envelope, and much shorter and
usually much narrower than the swollen part
of the cell in which it is formed. It is also
distinguished by the starch-reaction or
granulose-reaction described on page 18,
which is usually manifested · by the cells
before the spores are formed, and by its
habit, since it does not usually aggregate
and form distinct membranes or larger Zoogloeae, and at the
period of spore-formation often appears in the form of the motile
rods with capitate end, which were likewise noticed on a former
occasion (see page 17). Otherwise Bacillus Amylobacter is very
morphous; the most different special forms of sporogenous

Fig. 13.

Fig. 13. Bacillus Amylobacter. Motile rods, some cylindrical and
without spores, some swollen into various special shapes and with spore-
formation in the swelling. *s* mature spore with broad gelatinous envelope
and isolated by the deliquescence of their mother-cells. Magn. 600 times,
with the exception of *s* which is more highly magnified.

cells make their appearance irregularly mixed up together and connected with one another, as Fig. 13 shows.

In its mode of life, Bacillus Amylobacter is a type of Pasteur's anaerobia (page 54), though the possibility of its vegetating in the presence of oxygen is not excluded. Living in this manner it is first of all the chief promoter of butyric acid-fermentations of sugars, that is, of fermentations in which butyric acid is the primary product and is accompanied by other products which vary greatly according to the special material, as is shown by the researches of Fitz. It may also be assumed that it is this species which causes butyric acid-fermentation in lactates, though an objection brought by Fitz against this view has not yet been quite removed. In this special character of the ferment of butyric acid B. Amylobacter plays an important part in human economy, whether as the cause of fermentation in acid articles of vegetable food which in this case rapidly rot, or of the butyric acid-fermentation which is essential to the ripening of cheese.

Bacillus Amylobacter is also a specially active agent, as van Tieghem has shown, in the decomposition of decaying parts of plants by destroying the cellulose of the cell-membrane. It does not attack all cell-membranes, not for example suberised membranes, those of bast-fibres, of submerged water-plants, of Mosses and many Fungi; on the other hand, the membranes of fleshy and juicy tissues, as in leaves, herbaceous stems, cortex, tubers of land-plants, and softer kinds of wood, are especially liable to its attacks. In all these cases it first of all decomposes the cellulose into dextrin and glucose by means of a diastatic enzyme which it disengages, and these then undergo the butyric acid-fermentation. Most starch-grains escape its attacks, but paste and soluble starch are very liable to them. Hence the maceration and destruction of parts of plants that are kept wet are to a great extent the work of this Bacillus, alike in cases which involve the economical processes of mankind, such as the maceration and rotting in water of hemp, flax, and other textile plants, in order to obtain the fibres, and in such as the wet-rot of bad potatoes according to Reinke and Berthold. Van Tieghem is inclined to attribute to Bacillus Amylobacter a prominent part in the nutrition of ruminant animals, since it vegetates in their stomachs and splits up the cellulose of their food into soluble products of decomposition capable of re-sorption.

Van Tieghem has also shown or made it probable that this Bacillus has been an active destroyer of cellulose at least since the period of the coal-measures. Fossil plants silicified in a more or less advanced state of maceration, show in their sec-tions the same progression in the destruction of the cell-wall which is observed in macerated plants of the present day, and also the silicified remains of a Bacterium, which he identifies with B. Amylobacter.

The active powers of fermentation and decomposition of this Bacillus are not confined to the non-nitrogenous bodies just enumerated, as is shown by the investigations of Fitz, which have been before briefly mentioned. The details of these in-vestigations will be found in the works already cited. The behaviour of this Bacillus to proteids will be noticed presently.

Though there can be no doubt that much the larger number of fermentations producing butyric acid are caused by Bacillus Amylobacter, yet it cannot be said to be the exciting cause of all fermentations which have butyric acid for their primary pro-duct. On the contrary, Fitz describes a large round chain-forming Micrococcus and a short non-endosporous rod-shaped Bacterium, as ferments producing butyric acid in calcium lactate and in some sugars. His former statement, that Bacillus subtilis forms butyric acid by fermentation from starch-paste, and that this fermentation is a very advantageous method of procuring butyric acid, must be founded on a confusion of forms. The typical B. subtilis of Brefeld and Prazmowski can-not be the species intended, for Prazmowski distinctly states that it does not excite fermentation of any kind in starch-paste.

Vandevelde's observation (49) that B. subtilis certainly gives rise to slight fermentation in meat-extract, glycerine, and grape-sugar, after the oxygen is consumed, with special production of butyric acid, can hardly be taken into consideration, for he speaks of very small amounts produced from fermentation, while Fitz states that the amount of butyric acid produced is very large.

In the absence of more precise morphological observations, the species of the Bacillus subtilis of Fitz's starch-fermentation must for the present remain undetermined. In connection with this question I will here briefly repeat the remark made on page 74, that there are certainly several saprophytic and endo-sporous species of Bacteria, which are very like Bacillus subtilis, and have no doubt been frequently confounded with it. Nothing certain can at present be stated with regard to their effects as ferments. The B. subtilis of Brefeld and Prazmowski, the only form to which I give the name, is clearly distinguished from them by the assemblage of characters described in former lectures, and by the mode of germination (page 21), as well as by their collecting on the surface of the nutrient fluid into membranes which are folded in wrinkles, and consist of filaments disposed in parallel zigzags and ultimately forming spores, and by the ellipsoid comparatively broad spores themselves.

8. If we examine in conclusion the decompositions which occur in proteid compounds and in glue, we shall first of all see no reason to doubt that all of them, and especially those which are accompanied by the development of gas and are usually known as processes of putrefaction, are the work of Bacteria. The data which we possess show that the processes in these decompositions and the participation of the different species of Bacteria in them are, as might be expected, extremely multifarious. The work of distinguishing between the species of Bacteria concerned and their specific modes of operation is still in its infancy.

A point of great importance to notice here is the liquefaction of the gelatine which occurs in cultures of some Bacteria, for example Bacillus subtilis and B. Megaterium, but not in all.

And here the many-sided Bacillus Amylobacter must again be mentioned. According to the researches of Fitz and Hueppe, this Bacillus decomposes the casein of milk in the following manner : the casein first coagulates as when rennet is employed, the effect being produced by the enzyme disengaged by the Bacillus; it then becomes liquid and is converted into peptone, and then into further and simpler products of decomposition, among which leucin, tyrosin, and ultimately ammonia, have been ascertained. The fluid meanwhile acquires a more or less pronounced bitter taste. Similar if not identical effects on the casein of milk were observed by Duclaux to proceed from the Bacilli which he names Tyrothrix (see page 52), the greater part of which must also morphologically approach near to B. Amylobacter. Tyrothrix tenuis, for example, first produced the coagulation of rennet, then liquefaction, and after that formation of leucin, tyrosin, ammonium valerianate, and ammonium carbonate. There can be no doubt that these changes, and others connected with them, are the essential part of the phenomena which constitute the process of ripening of the cheese prepared from the coagulated milk, the above-named Bacteria, with some others, being contained in the cheese, and being procurable from it for the purpose of examination.

Bienstock (50) has recently submitted the Bacteria of human faeces to careful examination, and found that in those of adults, besides other forms which are unimportant in reference to the processes in question, there is always a particular Bacillus present, which he regards as the specific cause of putrefaction not only of the bodies contained in the faeces, but of those which contain albumin and fibrin. This Bacillus in a pure culture is able of itself to separate albumin or fibrin into the successive products of decomposition which have been ascertained in other cases of putrefaction up to the latest and final products, carbonic acid, water, and ammonia. If allowed to operate on one of the series of decomposition-products already prepared, tyrosin for example, it carries on the decomposition in the order of succession of the regular products of putrefaction. None of the other Bacteria examined by Bienstock produced these effects. Neither casein nor artificially prepared alkaline albuminates were rendered putrid by Bienstock's Bacillus; casein is even said to remain entirely unaltered; in accordance with this the Bacillus itself and the specific decomposition with the characteristic faecal smell are wanting in the intestine of sucking children.

As regards the morphological characters of this Bacillus of the decomposition of proteid, it appears from the descriptions of Bienstock that it is endosporous, and resembles B. Amylobacter in shape at least at the time of formation of spores, and like it forms the motile rods with capitate end (see page 17) which Bienstock compares with drumsticks. It is, however, smaller than B. Amylobacter, and even than B. subtilis. It is scarcely possible to form a clear idea of the course of development of this form from the investigations and descriptions which we possess, and we must wait for further enquiry into this point.

We must also wait to see whether the monopoly of putrefaction claimed for the drumstick Bacillus will be confirmed. This, with all due acknowledgment of the results as reported, is scarcely probable when we call to mind our experience of other processes of decomposition. I will not bring forward other accounts which have been given as arguments against the exclusive character of Bienstock's Bacillus, because precise discriminations of form are too recent to set aside the objection, that this Bacillus may be present though unrecognised where some other form is said to have been found, and may be the really active agent.

But it is necessary, at least, to refer in this place to these other statements, inasmuch as the view pretty generally entertained is that Bacterium Termo is the usual exciting cause of

putrefaction. Cohn expresses this opinion in the most decided
manner when he says that he has arrived at the conviction, that
Bacterium Termo is the ferment of putrefaction in the same
sense as beer-yeast is the ferment of alcoholic fermentation, that
no putrefaction begins without B. Termo, or proceeds without
its multiplication, and that B. Termo is the primary exciting
cause of putrefaction, the true saprogenous ferment. Although
we cannot now maintain these propositions in their full extent,
and although the expression putrefaction is employed in them
without any precise determination of the processes of decomposition and of the putrefying substance, yet on the one hand
there can be no doubt that the expression includes what is
commonly understood by the putrefaction of proteids, such
as meat, and on the other hand Bacterium Termo must,
at least, be a very constant attendant of such processes. It is
advisable, therefore, just to enquire what B. Termo actually is,
all the more since modern bacteriology itself scarcely ever uses
this old name. There is some ground for this, since it is
scarcely possible to make out with certainty what it was that
Dujardin, Ehrenberg, and others thirty years ago intended by
this name. What Cohn on the contrary described as B. Termo
in 1872 is an object as distinct as it is of frequent occurrence.
It is obtained by allowing the seeds, for instance, of leguminous
plants to rot in water, and then preparing a culture by introducing a drop of the putrid liquid into the solution known as
Cohn's nutrient solution for Bacteria[1]. The transference of a
drop of the solution thus infected several times in succession to
some fresh solution, sufficiently secures the purity of the culture.
The microscopic indication of the presence of Bacterium Termo
consists in the solution becoming more and more milky during
the first days of the culture, and then forming a greenish layer
on the surface, in which the form in question is collected in

---

[1] Cohn's normal solution, as given by Eidam in Cohn's Beitr. i. 3,
p. 210, consists of potassium phosphate 1 gr., magnesium sulphate 1 gr.,
neutral ammonium tartrate 2 gr., potassium chloride 0·1 gr., water, 200 gr.

extraordinary large quantities. Isolation by cultivation in gelatine is not possible, because the gelatine is at once liquefied by the rapid multiplication of the Bacteria.

Microscopic examination reveals a number of minute rod-like cells, according to Cohn's measurement about $1·5 \mu$ in length, and becoming one-half or one-third of that amount in breadth, engaged in active bipartition, and thus frequently united in pairs, but scarcely ever forming long rows; in this they resemble Micrococcus lacticus, but are distinguished from it by their somewhat larger dimensions, and especially by the very active independent movement of the individuals suspended in the fluid. The movement is often a peculiar backward movement in different directions. Zoogloeae are ultimately formed on the surface of the fluid in the form of greenish slimy films or lumps, in which the cells lie motionless. The alternation of these two states was clearly described by Cohn as long ago as 1853. Formation of spores in the characteristic manner has not been observed in B. Termo, and it must therefore be classed for the present with arthrosporous forms. I have said thus much in order to commend the old B. Termo to renewed observation; time will show how much will be left of it and its reputation as the exciting cause of putrefaction. I leave these sentences as they were originally written, only adding that Hauser has since shown that Cohn's Bacterium Termo is a collective species, and has resolved it into three kinds; but a closer comparison of these has still to be made (51). I now conclude with it the series of examples of saprophytic Bacteria.

X.

Parasitic Bacteria. The phenomena of parasitism.

WE now pass on to the second category of Bacteria, distinguished above on page 65 by their parasitic mode of life.

The term parasite is applied in biology to the living creatures which take up their abode on or in other living creatures, and feed on the substance of their bodies. The animal or plant which supplies food and lodging to a parasite is termed its host. Parasites are known in very different divisions of the animal and vegetable kingdoms, and many of them are well and certainly understood. I need only mention intestinal worms on the one hand, and on the other the long series of true Fungi which are parasitic especially on plants. Our experience of these forms, which are comparatively easy to examine, teaches us that the adaptations to the parasitic mode of life are extraordinarily complex and present an extreme variety of gradations between one case and another, that is between one species and another, and that these are dependent on the one hand upon the more or less strict requirements of the parasitic mode of life, and on the other upon the mutual relations between parasite and host.

To attempt to go at all at length into the above relations, would lead us here much too far into details. But we must call attention for our present guidance to one or two of the most important points.

As regards the nature of the parasitism, we have first the case which is farthest removed from the life of saprophytes, that of obligate parasites, which by the provisions of their nature can only complete the course of their development in the parasitic and not in the saprophytic mode of life. To take an example from amongst those with which we are best acquainted, this is true strictly and excluding all deviation into saprophytism of the Entozoa, such as tapeworms and Trichinae; among Fungi, of those which live inside plants and have been termed rusts (Uredineae). These organisms as a matter of fact live only inside their living hosts and feed on them. It is quite conceivable that the conditions necessary for their development may arise or be artificially produced outside the living host, and it would certainly be an instructive experiment to grow a tapeworm from the ovum in a nutrient solution; but this has never been actually

done, and no instance of the kind is to be found in nature.   In
these cases the parasitism is obligate and indeed strictly obligate.

I add the word strictly, because there is a modification of
obligate parasitism, in which a parasitic mode of life is necessary
for the completion of the entire course of the development, and
is often the only one which actually occurs, while at the same time
saprophytism may take its place, at least in certain stages of the
development.   No example of this kind occurs to me at this
moment from the animal kingdom, but there are some to be
found.   Among the Fungi there are a number of species of the
genus Cordyceps which inhabit insects, especially caterpillars,
and in which this adaptation occurs in a marked manner.   The
germ-tubes developed from spores on the caterpillar penetrate
into the insect, spreading luxuriantly in it, and at length killing
it, and after its death they fill the whole of its body with mycelial
tissue.   From this tissue, if the conditions are favourable for
vegetation, large Fungus-bodies are produced, several inches in
length, which are the stromata of the Fungus, and produce spores
(ascospores).   These go through the same course of development,
if they also find their way to a suitable living insect.   But if this
does not happen, the spores have the power of germinating on
a dead organic substance, for instance in a nutrient solution,
and their germ-tubes may develope there into Fungus-plants.
But these plants do not produce the characteristic stromata just
mentioned.   They form different spores from those produced
in the stromata, and these spores may also develope in the
saprophytic mode of life; but if they find their way to the
proper insect-host, they can recommence the course of development which reaches its highest point in the formation of stromata   as   described   above.   Here   then   we   have   parasites
which are able to complete a certain portion of the course of
their development while living as saprophytes, though without
reaching its highest point, namely the formation of stromata;
they may be shortly termed facultative saprophytes.

Thirdly, there are also facultative parasites.   These are species which are able to develope as perfectly, or at least nearly as perfectly, in the saprophytic as in the parasitic mode of life. The ' or ' shows at once that there are gradations also within this category, and these are, as might be expected, of such a kind that some plants find the conditions more favourable in the parasitic, others in the saprophytic mode of life, while others again show no difference in this respect. There are many instances of these modifications of facultative parasitism among the Fungi, and we shall soon make acquaintance with similar instances in the Bacteria.

The mutual relations in each case between parasite and host, the dependence of the one on the other, the benefit or injury which the one receives from the other, are independent of these strict requirements of parasitism which vary in each separate case. In cases like that of the Trichinae, for example, we are in the habit of speaking of this relation as one-sided, as one in which the parasite derives its entire means of subsistence from the living host, while the host receives nothing but harm from the parasite through the necessary withdrawal of its substance and other manifold chemical and mechanical disturbances which it suffers. States of disturbance of the normal existence of a living being, the normal requiring to be determined in each case by experience, are known as diseases ; the parasites of which we are speaking cause these states and are therefore injurious to health, the exciting causes of disease. Further, the parasite by means of its germs, spores, ova, or whatever other name is given to its organs of propagation, may be transferred from the host which has been made ill by it to others in which it will also produce disease. The maladies caused by parasites are therefore transferable from host to host, they are, to use the common expression, infectious.

But these cases, in which the parasite is injurious to health and the injury is all on one side, are only one extreme among those that are known. There are others in which the two organisms live in common with equal advantage to both, and there is every possible gradation again between these extremes. Lastly, there are cases in which a parasite lives in a host without either injuring or sensibly profiting by it, at most deriving its food from the refuse of the metabolism of the host. In extreme cases of this kind, which obviously lie on the border-line of the phenomena of true parasitism, we use the expression lodger-parasites.

Further, there is a fact more or less known to the experience of every one, which holds good of all the categories of parasites distinguishable according to the points of view here indicated, namely that a parasitic species may make choice as we should say between the hosts which it occupies, that is, attacks one host and thrives perfectly well in or on it, while it either refuses others altogether or at least grows less vigorously in them. In these respects also there is every conceivable gradation. First, as regards the choice of the host-species by a parasitic species; one extreme is marked by the narrowest one-sidedness. For instance, a strictly obligate and very well marked parasitic Fungus, Laboulbenia Muscae, mentioned on page 39, grows exclusively on the house-fly and on no other insect, at least according to our present investigations. Other Fungi and other parasites besides the Fungi are not so one-sided in their choice, since they attack a larger number of host-species, but those only as a rule which belong to a narrow cycle of affinity, a genus, family, &c. Thus, for example, some of the species of Cordyceps mentioned above, grow in the larvae of a great variety of butterflies and other insects. But it sometimes happens that single host-species within such a cycle of affinity remain excluded from the choice for reasons of which we are ignorant. Lastly, we are acquainted with obligate and facultative parasites which are able to complete their development equally well in hosts of the most different cycles of affinity. I need only mention Trichina spiralis once more, which thrives well in rodents, swine, men, and other animals. Examples in plenty may also be drawn from the Fungi; but among them also strange exceptions occur within a cycle of preference, some host-species being spared by the parasites without any definable reason.   To name only one instance, a Fungus named Phytophthora omnivora from its catholicity of taste attacks the most heterogeneous plants, species of Oenothera and other herbs and garden-flowers, Sempervivum, the beech, &c.; on the other hand it never attacks the potato-plant, which its nearest relative Phytophthora infestans prefers.

It is at present scarcely possible to give an exact account of the physiological causes of these preferences, but it is at the same time obvious that they depend essentially on chemical and physical qualities and distinctions.

If then there is a choice between one species and another, there must be a similar choice to a certain extent between individuals of the species, for the differences between the several species are the same in kind, though not in degree, as those between individuals of one and the same species; the latter are less than the former, and are therefore also less pronounced, being sometimes scarcely or not at all perceivable; gradations between the different cases which we meet with everywhere are also not wanting here.

If we describe these phenomena which run through the whole of the long series of parasites in the reverse way, that is to say, not with reference to the parasite but to the host, then we say that the host is differently suited, disposed, or predisposed, according to the species and individual for the attack of a parasite. We may speak of predisposition on the part of a species, or individual, or in different states, stages of development, or ages of an individual.   Of these individual predispositions it may be further specially observed that they must, no less than all others, as a general rule, have their foundation in each case in the chemical, physical, and anatomical constitution.   It may be shown for example in the case of certain Fungi of the genera Pythium, Sclerotinia, and others which live in plants, that individuals of the same host-species have unequal susceptibility to the attacks of the parasite, and unequal power of resisting them according to the relative amount of water which they contain. Since in these cases younger plants have more water than the older, a predisposition dependent on age is also indicated accordingly.

In cases where the parasite causes a disturbance, which we call sickness, of what by experience is regarded as the normal vegetation of the host, if the predisposition is individual we speak usually of a sickly predisposition. This may be correct, in so far as the predisposition to the attack of the parasite may be connected with deviations from the state which is from experience termed the sound state. But it need not always be correct, for there is no reason at all why the disposition for the attack of the parasite should in every case indicate a condition, which must be called sickly even when there is no parasite present. The above-mentioned example of the predisposition varying with the age is a sufficient proof of this. Here we must distinguish between one case and another, and care is required in determining each individual case.

An example may help to make this still more plain; it is that of a case which is comparatively very accurately known. The common garden-cress, Lepidium sativum, is often attacked by a parasitic Fungus of comparatively large size, Cystopus candidus. In consequence of this it shows considerable degeneration, swellings, curvatures of the stem, and often also of the fruits, and on these parts and on the leaves white spots and pustules subsequently turning to dust, which are formed by the sporogenous organs of the Cystopus, and give the entire phenomenon the name of the white rust in cress. This is a case of disease, and so striking that every one notices it at once with the naked eye. Now we find in a bed of cress at about flowering time a certain number of rusty plants, two for example or twenty. They are in the middle of the other hundred or thousand plants, and these are healthy and free from the Fungus and continue so till the period of vegetation is at an end. This is the case, though the Cystopus forms countless spores in the

I

white rust-pustules, and the spores are dispersed as dust and are at once capable of germination, finding the necessary conditions for their further development in the bed of cress, and are the instruments by which the white rust-disease is eminently infectious.   Nevertheless those hundred or thousand plants are not infected. All that has been hitherto said is strictly correct, and if we limit our view to this, we shall see in the phenomena which have been described a conspicuous case of individual difference in predisposition; a case too perhaps, if we judge hastily, of sickly predisposition in the plants attacked, for they do become sick and the others do not.   And yet this is not the true account of the matter.   Every healthy cress-plant is equally liable to the attacks of the Cystopus and to the rust-disease which it causes, only the liability is confined to a certain stage of the development, and ceases once for all when that stage is past.   The germinating cress-plant in effect, first unfolds two small three-lobed leaves, the seed-leaves or cotyledons.   When it has grown a little further and formed more foliage-leaves, the cotyledons wither and drop off.   It appears then, that the germ-tubes of the Fungus of white rust find their way into all the cotyledons and are able to develope there, and if this development has once begun, the Fungus establishes itself at once in the tissue into which it has penetrated, and grows on in and with the growing plant, and produces the disease.   The germ-tubes of Cystopus may indeed make their way for a short distance into all the other parts of the plant, but are unable to establish themselves inside it and continue their development.   The plant is for the future safe from the attacks of the parasite as soon as the cotyledons have fallen off.   The two or the twenty rusted plants in the bed are the ones in which the Fungus attacked the cotyledons in good time; if it had attacked the thousand others in equally good time, all would have been rusted.   They continued healthy, because they were not infected in the stage in which they were open to infection, that is, predisposed.

It follows necessarily from what has been said with respect to the variety of gradations in the mutual relations of host and parasite, that the progress and issue of the disease must also show manifold gradations, varying with the species on both sides, and in a less degree with the individuals. The very general occurrence of Trichinae, tapeworms, the itch and other diseases bring them so much under the notice of all, that this brief notice of them here will be sufficient.

XI.

Harmless parasites of warm-blooded animals. Parasites of the intestinal canal. Sarcina. Leptothrix, Micrococci, Spirillum, Comma-bacillus of the mucous membrane of the mouth.

IT seemed to me expedient to give the foregoing short review of the phenomena of parasitism and of its consequences, because all that we know of parasitic Bacteria are only special cases of the main phenomena which occur everywhere; and this is equally true of all our suppositions concerning them. The understanding of these matters will therefore be materially aided by resting on old and long-known phenomena.

In passing now to the consideration of important examples of parasitic Bacteria, it will be advisable to speak first and chiefly of the parasites of warm-blooded animals, including mankind, and afterwards to say a few words on those of other animals, and of plants.

In the former class it will best suit our purpose to distinguish the species which are the exciting causes of disease from those which are less injurious or altogether harmless; and first a few words respecting the latter kind.

The digestive canal and the breathing passages, the former especially, are a favourite habitat of lower organisms, Fungi and Bacteria, putting the various kinds of worms out of consideration. A large number of Fungi make use of the intestinal canal

as a regular, if not in most cases an absolutely indispensable
thoroughfare, since when introduced with the food they find a
home and nourishment in it for the first stages of their development, and then complete it on the voided faeces. This is shown
by the abundant and remarkable Fungus-flora of dung.

It is known that many forms of Bacteria occur in large
numbers in the contents of the intestinal canal. A more
thorough sifting and sorting of most of the species has yet to be
undertaken. In the human intestine Nothnagel has distinguished
Bacillus subtilis, B. Amylobacter, and other not clearly defined
forms, and Bienstock (50) his drum-stick Bacillus. Kurth found
his Bacterium Zopfii in the intestine of fowls (see page 22). To
these examples must be added the constant, and according to
van Tieghem (see page 102), the essential presence of Bacillus
Amylobacter in the stomachs of ruminants (52).

The acid of the gastric juice may prevent the appearance of
most of the Bacteria in the normal contents of the stomach (in
the rennet-stomach in ruminants). Koch's researches into
anthrax, to be noticed again presently, have even shown that
Bacillus Anthracis in the vegetative states is killed by the gastric
juice, and only its spores maintain their vitality. This may be
the case with some other species, and it may be of some importance that a kind of sorting thus takes place in the normal
stomach, by means of which some only of the Bacteria introduced with the food reach the intestinal canal in the living state.

That the gastric juice has not always an injurious effect, but
that here too there is a difference between one case and another,
is shown by the researches of Miller and W. de Bary (52). We
are acquainted with one species, the well-known Sarcina ventriculi (Fig. 14), which thrives particularly well in the human
stomach. This species forms packets in the shape of almost
perfect cubes of roundish cells arranged in regular layers
parallel with the surfaces of the cube, and kept firmly united by
tough gelatinous membranes. Comparison shows plainly that
the packets are formed in the manner which can be directly

observed in the case of other very similar species (see Fig. 15), namely, from a single round initial cell by successive divisions formed in three directions. The packets separate as they grow into daughter-packets, each of which contains the progeny of one of the cells of previous orders of division, and as this process is repeated the packets multiply. Nothing further is known of the history of the development of Sarcina.

Sarcina ventriculi is at present known only from the human stomach and intestinal canal. In diseases, especially enlargements, of the stomach, it is often found in incredible quantities.

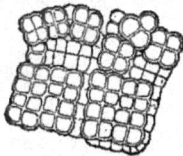

Fig. 14.

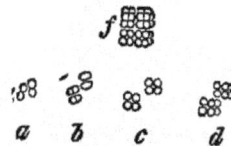

Fig. 15.

Yet no causal connection has been ascertained between its occurrence and distinct phenomena of disease, and other conditions being the same it may be present in profusion or sparingly, or be absent; its absence indeed is the rule in much the larger number of stomachs, diseased as well as healthy. The causes of all this are unknown; nor can we tell whence it finds its way into the stomach. Its occurrence outside the stomach

Fig. 14. Sarcina ventriculi, Goodsir. Large-celled form just taken from the contents of the stomach of a patient and imbedded in soft gelatine; a comparatively small and cube-shaped packet. View of one surface only; but other surfaces project below and to the right beyond the edge of the first. In the surface depicted the cells with double contour-lines are rightly focussed; those with single contour-lines are not in the right focus but lying at a lower level. Magn. 600 times.

Fig. 15. Sarcina minuta, de Bary, in gelatine on a microscopic slide. *a-d* successive states of the same specimen, observed as a double pair of round cells, *a* about 4, *b* about 6, *c* about 9, and *d* at 10 o'clock in the afternoon. In *c* the tetrads are still formed of a single layer, in *d* a division has taken place in each cell in the plane of the paper; each tetrad developes into an 8-celled cube-shaped packet. *f* a 32-celled pocket. See note 53.

and intestine, excepting of course in the evacuations, has not
been observed with any certainty, and the attempts to cultivate
it have, up to the present time, been without success in all cases
offering clear results.

It is true that we are acquainted with a number of forms or
species, in which the cubical packets are so like those of Sarcina
ventriculi that they must be placed alongside of it as closely
allied forms.   These occur both outside living organisms, as
saprophytes, and also in the bodies of living animals, and
among them of men.   That they are not very widely diffused
is evident from the fact that the reported cases of their occurrence are always solitary ones.

Saprophytic forms of Sarcina have been found casually, that
is without having been introduced intentionally, by Cohn and
Pasteur on all kinds of nutrient solutions, by Schröter on boiled
potatoes, by myself on acetified beer, on coagulated milk, and
elsewhere.   In these instances the yellow forms (Sarcina lutea,
S. flava) have repeatedly been observed (53).

Sarcina-forms inhabiting the bodies of living animals are
described as obtained from the bladder (S. Welckeri), the lung,
the mouth, and other cavities of the body, even from the blood
of the human subject, and from the cavities and the intestinal
canal of other warm-blooded animals.

These forms, the saprophytic as well as the parasitic, are, so
far as the statements before us enable us to judge, without
doubt clearly distinct from Sarcina ventriculi.   Unfortunately
many accounts are so defective, so very much restricted, one
might say, to the word Sarcina, that it is impossible to determine
their identity.   The fact moreover which was noticed in the
case of Sarcina ventriculi is also true of the parasitic forms ;
their occurrence, as far as our knowledge goes, has not been
shown to be in causal connection with distinctly morbid phenomena, and they must for the present be regarded as simply
lodger-parasites (53).

Many kinds of Bacteria are observed in the mucous membrane of the mouth and nose. With regard to the latter this assertion favours the supposition that Bacteria are uniformly present in that catarrh of early summer which goes by the name of hay-fever. I can bear out this statement myself as a sufferer from this disagreeable malady, though I must add that Bacteria are also present during the 10–11 months of the year that are free from hay-fever. I found them to be small, short rods resembling those of Bacterium Termo. Whether specifically different forms are present, or predominate at different times, has not been ascertained.

We are better acquainted with the abundant growth of Bacteria in the mucous membrane of the mouth. They occur in greatest profusion on the gums and between and on the teeth, in a more scattered manner, but still in considerable numbers on the rest of the surface of the mouth and in the discharged saliva. A specimen of mucus scraped from a tooth is seen to be chiefly composed of a form known by the old name of Leptothrix buccalis, Robin (Fig. 16, *a*). It consists of long straight filaments glued together into dense bundles, brittle and readily separating into pieces transversely, and of unequal thickness; larger filaments

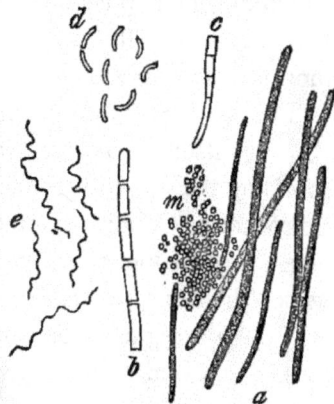

Fig. 16.

Fig. 16. Bacteria from mucus of the teeth. *a* Leptothrix buccalis, Robin; filaments or portions of filaments of different thickness. *b* a portion of a filament after treatment with alcoholic solution of iodine showing the segmentation distinctly. *c* portion of a filament much narrowed at one end, without treatment with reagents and showing segmentation distinctly. *d* Lewis' comma-bacillus, that is, a short-celled Spirillum. *e* Spirochaete Cohnii, Winter (Spirochaete of mucus of the teeth, Cohn, Beitr. i. 2, p. 180, and ii. p. 421). *m* Micrococcus-heaps. All the figures are of specimens from the same preparation, *e* and *b* after staining, the rest fresh from the mouth. Magn. 600 times, with the exception of *b*, which is more highly magnified.

are over 1 μ in the transverse diameter, others only half that thickness. The length also of the members (cells) is unequal, in some cases not exceeding the transverse diameter, in others several times greater. The filaments, especially those with short and thick cells, often show the reaction of granulose (page 5), but different portions of the same filament may assume alternately a blue and yellow colour with iodine. Rasmussen (54) claims to have distinguished three separate forms of Leptothrix buccalis by aid of cultivation. I cannot say whether this is rightly done or not, for Rasmussen's work is only known to me at second hand.

Secondly, the masses of Leptothrix often contain round Cocci, which are sometimes irregularly rolled up into dense gelatinous heaps, and like the Leptothrix-forms are without the power of motion (Fig. 16, *m*).

Thirdly, a Spirillum-form is commonly found with the others, showing itself rather in single specimens and in the fluid surrounding the Leptothrix-masses after addition of water; this is Spirochaete Cohnii, Winter (S. buccalis or S. dentium), and consists of filaments of extreme tenuity without evident transverse divisions, spirally twisted into 3–6 or more steep and often irregular coils, flexible and either exhibiting a slow twisting movement or else without motion (Fig. 16, *e*).

Lastly, one more form is often though not always observed with the others, a thin short rod-shaped Bacterium, bent like a bow, and described first by Miller and then by Lewis (55) as the comma-bacillus of the mucus of the mouth (Fig. 16, *d*); in a fluid it usually exhibits an active hopping movement.

It may be assumed beforehand as certain, that besides these forms other saprophytic Bacteria must also occur in the mucus of the mouth. Miller, according to recent communications, has found twenty-five such organisms. Hueppe speaks of two Micrococci which produce lactic acid as coming from the human mouth (see page 94). But other forms do not appear to be developed in any abundance in healthy individuals. It may be said perhaps that their invasion is hindered by the presence of the characteristic dwellers in the mouth mentioned above.

I repeat that I would have these latter spoken of at present only as forms which are actually present side by side, nor will I enter further into the question how far they stand in genetic relation to one another. From the impression which they give us and from present investigations it seems to be highly probable that we have before us several distinct social species.

The dwellers in the digestive and respiratory passages which have now been described, together with other near allies found in mammals are, so far as our knowledge goes, almost without exception harmless guests, lodger-parasites only, those which live in the mouth being perhaps even useful as protectors against an invasion of destructive ferment-forms. Certain forms however are disagreeable exceptions, inasmuch as they cause caries, the disease in which the teeth become hollow. Every hollow tooth is penetrated throughout by Bacteria, and different forms or species occur in different cases; Miller (55), after examining hundreds of teeth, has distinguished five of these forms. He has also shown by very thorough investigation that one of them, a Micrococcus, forms lactic acid in a substratum containing sugar or starch. The salts of calcium in the substance of the tooth are dissolved by the excreted acid, and the Bacterium is thus enabled to force its way into the tooth; as more and more of the calcium is withdrawn the Bacterium passes into the tubuli of the dentine of the tooth, and ultimately spreads through and destroys the tooth. It can scarcely be doubted that Miller's four other species produce the same effects.

XII.

Anthrax and Fowl-cholera.

LEPTOTHRIX buccalis, as the exciting cause of caries in the teeth, carries us on to those parasites in warm-blooded animals which produce disease.

The best way to obtain a clear idea of these organisms, their manner of life, and their effects will be to examine first of all some comparatively well-known examples.

Let us take first the disease known as anthrax, charbon, sang de rate and its exciting cause, Bacillus Anthracis (56).

Bacillus Anthracis has already been repeatedly brought before our notice. Its description will therefore only be briefly recapitulated here, and its figure reproduced (Figs. 17, 18). It consists of cylindrical cells about 1–1·5 μ in thickness, and 3–4 times that length. In the blood of animals these cells are usually connected together into long straight rods (Fig. 17, *c*), which appear homogeneous till they are carefully examined, that is, do not distinctly show segmentation into individual members. When grown in a dead substratum the rods develope into very long filaments, which appear sharply bent in several places, and form curvatures and loops; they also separate at the points of flexure into rod-shaped pieces, and are usually collected in large numbers into bundles or sheaves and twisted round one another (Fig. 17,*a*). The rods and filaments are without the power of locomotion, except in special cases, which will be noticed below. The formation and germination of the spores take place in the manner described in Lecture III in the case of the endosporous Bacilli; in germination the spores merely grow in length (Fig. 17, *b*) without throwing off any distinct spore-membrane, and the young germ-rod often exhibits a slow oscillating movement. The ripe spore is broadly ellipsoidal, as broad as its mother-cell which retains its cylindrical shape but much shorter, and lies very nearly in the middle of the mother-cell till it is set at liberty by the swelling of the membrane.

Bacillus Anthracis is distinguished under the microscope by the absence of independent motion and by the form of its germination from B. subtilis, which is very much like it in almost all respects, except that it is usually more slender and is not parasitic. To these must be added in ordinary cases the macroscopic distinction, that B. Anthracis at the highest point of its development forms a floccose sediment in nutrient solutions, while B. subtilis forms the dry membrane on the surface mentioned on page 12. Some exceptional phenomena will be noticed further on.

Anthrax chiefly attacks mammals, and among these the most liable to it are herbivorous animals, especially rodents and ruminants; of the species which have been observed, domestic mice, guinea-pigs, rabbits, sheep, and horned cattle are the most susceptible in the order of naming. The next in the degree of liability are omnivorous animals, and among them men; then

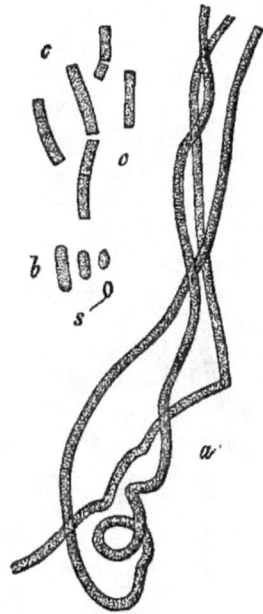

Fig. 17.

come the carnivores, and among these cats, for instance, are more frequently attacked than dogs. Birds also are said to be susceptible to the disease, though this has been disputed; Gibier found that frogs, and Metschnikoff that lizards (Lacerta viridis) were susceptible when they were kept at about the temperature of the bodies of warm-blooded animals. We will pass over

Fig. 17. Bacillus Anthracis. *a* portion of a group of vigorously growing filaments; the segmentation into cells is not visible, but has nevertheless taken place. *b* three successive stages of a germinating spore; close by is a ripe spore *s* before germination. *c* rod from the blood of an infected guinea-pig some hours after the death of the animal, after treatment with distilled water. *a* and *b* from cultures on microscope-slides in solution of extract of meat. Magn. 600–700 times.

disputed points on the present occasion, leaving them to be examined in the special literature of the subject (56), and confine ourselves to cases which have been certainly ascertained, especially those in mammals. Susceptibility to infection varies with the species, as appears from what has been already said; within the limits of the species it varies according to race and age, and with the individual.

Anthrax is a widely-spread form of disease; at the same time it is matter of long experience, that it appears with unusual frequency in certain districts, and that these anthrax-districts are especially dangerous to herds of cattle, and are therefore dreaded by breeders.

The clinical aspect of the disease is different in different species of animals; in larger ones it is said to run a comparatively slow course, being accompanied with violent fever, &c., and in most cases but not always ends in death. Mice and guinea-pigs succumbed to the disease almost without exception in the cases observed, but without showing any particularly striking symptoms

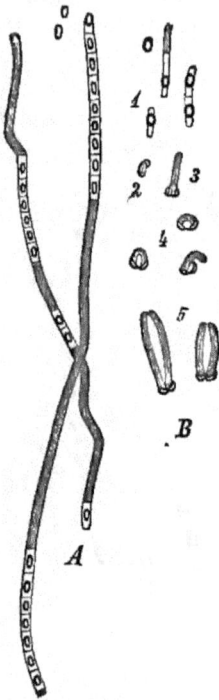

Fig. 18.

Fig. 18. *A* Bacillus Antbracis. Two filaments partly in an advanced stage of spore-formation; above them two ripe spores escaped from the cells. From a culture on a microscope-slide in a solution of meat-extract. The spores are drawn a little too narrow; they are nearly as broad as the breadth of the mother-cell. *B* Bacillus subtilis. 1 fragments of filaments with ripe spores. 2 commencement of germination of spore; the outer wall torn transversely. 3 young rod projecting from the spore in the usual transverse position. 4 germ-rods bent into the shape of a horse-shoe, one subsequently with one extremity released. 5 germ-rods already grown to a considerable size, but with both extremities still fixed in the spore membrane. All magn. 600 times.

up to the moment of death. In many instances I observed guinea-pigs lively and eager for food, till they all at once (about forty-eight hours after infection) collapsed and died after a short struggle.

If a diseased animal is examined a little before or immediately after death, the vegetative rods of Bacillus Anthracis (Fig. 17, *c*) are found in its blood. In larger animals, such as horned cattle, their numbers appear from the accounts before us to vary in every case. I say appear for reasons to be given presently. But they are always found in the capillaries of the internal organs, at least in the spleen. Koch states that they are not numerous in the blood of rabbits and mice, but are all the more numerous in the lymphatic glands and in the spleen. In guinea-pigs, the animals which I have myself chiefly examined, the entire mass of the blood is permeated by the rods; the smallest drop of blood, scarcely visible to the naked eye, taken from a slight puncture in the ear, or toe, or elsewhere, contains them; they are present in enormous numbers in the small vessels and capillaries of the liver, kidneys, spleen, and other organs. The same state of things continues for some time also after death; at a later period, when the first rigidity of death is passed away it often changes visibly, and it is possible to obtain considerable quantities of blood from the large blood-vessels, or from the heart of the animal without discovering a single rod in them. The rods are there but they are inclosed, often in large numbers, in the clots of fibrin, which it may be said in passing will supply the purest material for the culture of the Bacillus. It is quite possible that in the cases in which few rods were observed, the reason was, that those which were inclosed in the clots of fibrin were overlooked when the dead animal was examined after the coagulation of the blood; this is a point to be remembered in connection with the statements mentioned above, that the Bacilli are present sometimes in larger sometimes in smaller quantities.

The rods were first seen by Rayer in 1850, and next by Pollender, independently of Rayer, in 1855. The causal connection between the Bacillus to which the rods belong and the disease known as anthrax was first distinctly pointed out by Davaine in 1863, and this view of the matter, though it has been opposed on various grounds, is now accepted. It has been distinctly proved that the disease makes its appearance only when the Bacillus has found its way into the blood, and that the artificial introduction of the Bacillus into the blood results in the characteristic infection, the sickening of the animal. The infection follows if the living Bacillus is introduced directly into the blood by intentional inoculation, anthrax by inoculation, or unintentionally through wounds of the skin, anthrax by wounds, or through the uninjured mucous membrane of the intestinal canal, intestinal anthrax. It is effected both by means of living rods, and by spores, the latter then germinating in the blood or in the intestine. In both cases it is a matter of indifference whether the matter used for infection is obtained direct from a diseased animal or from a culture, such as will be described presently, which has been kept free from every trace of any product of animal disease. The Bacillus, when it has died of itself or been killed, is incapable of producing the infection.

When once the Bacillus has found its way into the blood of an animal capable of the infection, it grows and multiplies in the rod-form and spreads partly by its own growth, partly by the movement of the blood which carries the rods along with it in the manner described above. In proportion as this takes place, the sickness increases till at length death supervenes. The minutest possible quantity of the living Bacillus is sufficient to set these processes going. A guinea-pig, for example, dies with the symptoms which have been described in forty-eight hours after a quantity of spores or rods, too small to be visible with a pocket-lense, is introduced into the skin on the point of a needle by a puncture too slight to draw blood.

Anthrax by inoculation and anthrax by a wound are alike caused by the introduction of both spores and living rods.

Intestinal anthrax on the contrary is actually produced only by the introduction of spores into the body, as has been proved by Koch and his colleagues. In the natural course of things the Bacillus can only reach the mucous membrane of the intestine from the mouth, that is, when it is swallowed with food. It has then to pass through the stomach, and here the rods lose their efficiency, doubtless from the effects of the acid gastric juice; whether they are actually killed by it I am not prepared to say. The spores, on the contrary, pass unaltered through the stomach; they find the conditions favourable to germination in the contents of the intestine, and the rods developed in germination make their way into the mucous membrane of the intestine, especially through the lymph-follicles and Peyer's patches; from there the way is open through the capillary vessels in the mucous membrane to the passages of the blood.

It appears from the investigations of the above-named ob- servers that animals which chew the cud are liable to be infected in this way from the intestine. The experiments were made with sheep. Experience also with horned cattle not purposely infected seems to show by its agreement with these experiments that the latter animals also are liable to this mode of infection. It gives also the important practical result, that the cases of anthrax which occur in them spontaneously, that is not inten- tionally produced for experiments' sake, are chiefly of the intestinal form, and are caused therefore by taking in the spores of the Bacillus with the food.

Other animals are less susceptible of infection from the intes- tine; yet some attempts made to procure it succeeded in the case of guinea-pigs, rabbits, and mice; they were all unsuccessful with rats, fowls, and pigeons.

After all these experiences the first question is, whence do the spores come from which enter the animal? They are not formed either in the living animal or in the unopened carcase, for there vegetative development only takes place. But it was shown in a former Lecture that the Bacillus can not only germinate and vegetate luxuriantly outside the body of the animal, but that it forms its spores, in great profusion if the conditions are favourable, only outside the animal. The conditions for this non-parasitic development are the same as the general conditions given above for saprophytes. A supply of oxygen is required for perfect development; the optimum temperature for the formation of spores is 20–25° C. A great variety of organic bodies can serve as nutrient material, as experiments show, not only such as are of animal origin, for instance, portions of an animal that has died of anthrax, or the often bloody evacuations of diseased animals, or the meat-extract-solution in which the culture of the Bacillus was first accomplished (see page 12), but also many different parts of plants if they are not too acid, such as potatoes, turnips, seeds, &c. On the moist surface of such objects the Bacillus grows and forms extensive membranous coverings, which at the close of their vegetation produce countless spores.

Thus it is evident that the Bacillus of anthrax belongs to the class of facultative parasites, as described on pages 109, 110. It is above all a saprophyte, for it is not only able to prolong its existence in the saprophytic mode of life, but it requires it in order to attain to the highest phase of its development, the formation of spores. On the other hand it is capable of parasitism, since it finds its way into the proper host, and there acts as the exciting cause of disease in the manner which has just been described.

The phenomena attending the appearance of anthrax are now completely explained in all essential points from the mode of life of the Bacillus, if we take its existence for granted in the same sense as that of any other animal or vegetable species. The fact that anthrax when spontaneous usually appears in the intestinal form, shows according to the knowledge which we have acquired of it, that it passes in the spore-form from the saprophytic into the parasitic state, and that the route which it takes must be the same as that which serves for the food taken in by the animal. The starting-points in this migration must then be the places where the fodder is produced, meadows, grazing-grounds, &c. It is obvious that the Bacillus finds opportunity for its vegetation, and in the heat of the summer season the requisite temperature for forming its spores, on the dead organic bodies which are always to be found in these localities, and that when it has once established itself it can pass the winter there (see above, page 51), and thus remain from year to year in readiness to establish itself in a suitable host.

It is more difficult to say precisely why certain localities are the favourite homes of anthrax whilst others are free from it. Koch has shown some ground for thinking that the preference may be connected with conditions of moisture and inundation, so far as these affect the vegetation and diffusion of the Bacillus. I have not the requisite material for forming a decided opinion. It follows from what has been before said that the Bacillus is not obliged to return from its existence as a parasite and from the body of the attacked or dead animal to saprophytic vegetation in the infected places; as it is not obliged to pass through the state of parasitism, it can, as we learn from cultures, live as a saprophyte through an unlimited number of generations. On the other hand experience as certainly shows that it can return from the sick or dead animal to the life of saprophytism, for it remains in the animal long after its death and continues alive and capable of growth, and it can in fact return to the ground and to the condition of a saprophyte with the bloody dejecta, which, as we are told, larger animals void in a severe attack of anthrax, or with the decomposing carcases and the substances flowing out therefrom, which are excellent food for them.

The Bacillus can at the same time also be transported as a parasite from one place to another, and the spots on which sick animals fall down or their dead bodies are buried may become the abodes of the disease, as experience has long since shown. For the same reason a locality may possibly continue perma-

K

nently infected. If no herds of cattle visit them, small animals,
the rodents especially, which are so susceptible of the disease,
may ensure the introduction and conservation of the Bacilli.
But these circumstances are, as was said, not directly necessary
for the existence of the Bacillus and of the risk of anthrax, notwithstanding the importance which has been attributed to some
of the conditions connected with them.

To complete this review of the subject we may remark in
conclusion, that the Bacillus and its effects appear to be transferable as might be expected from one living animal to another,
and the disease is propagated by this transmission.   This infection belongs of course to the category above distinguished, of
anthrax as produced by inoculation or by a wound.   It can
only be brought about by means of vegetating rods, because
these alone are present in the living animal, and the rods, as we
saw, must be conveyed directly into the blood of the living
animal, to be capable of further development.   And now the
conditions of infection have been sufficiently indicated for
our purpose. A possibly exaggerated importance has been
ascribed to stinging flies and gnats as agents of infection,
since these insects, when they have sucked blood from an
animal which contains Bacilli and then puncture a healthy
animal for the same purpose, may easily effect a true inoculation
of the disease.

The Bacillus of anthrax in the character of a parasite produces in the animals and in the cases which have been described
injurious effects, which may be provisionally compared to those
of a poison, and may therefore be termed poisonous and
virulent.

This virulence may be gradually attenuated till it ceases to be
dangerous even in the case of that most susceptible of all the
animals experimented on, the domestic mouse. Pasteur has
shown that this takes place when the Bacillus is cultivated in a
neutral nutrient solution, meat-broth, especially chicken-broth
with a plentiful supply of oxygen at a temperature of 42–43° C.

Toussaint and Chauveau obtained the same results at a higher temperature. A culture of this kind ends in the death of the Bacillus, which ensues, according to Pasteur's account, in about a month or a little more than a month. Till this time the Bacillus vegetates without change in its morphological characters, except that the formation of spores is delayed or altogether stopped. That it never takes place is denied by Koch, Gaffky, and Löffler on the ground of direct observation. If the Bacillus is transferred to a fresh culture at any time before death has supervened, it will develope in the normal manner, and even produce normal spores at the proper temperature. If the temperature continues to be raised, complete attenuation results in a shorter time; a few days are required at a temperature of 45° C., a few hours at one of 47° C., a few minutes at one of 50–53° C. The three Berlin observers found that there was a considerable difference between 42° and 43° C. in the time required for total attenuation in correspondence with the temperature-differences of tenths of a degree, the process being accelerated as the temperature rose.

It follows that if the Bacillus is cultivated at a temperature between 42° and 43° C., we obtain material which will be harmless to animals in the order of their susceptibility to infection, for instance to rabbits first, after them to guinea-pigs, and lastly to mice; fluctuations will naturally occur according to individual susceptibility, age, and other circumstances.

It has been already stated that the Bacilli are capable of further vegetation after passing through every degree of attenuation of their virulence before reaching the stage of actual loss of vitality. If their cultivation is continued under optimum conditions they grow in their normal shape and form normal spores, but the successive generations, even when produced from spores, nevertheless retain as a rule the degree of attenuation of the first generation; some kill mice for example and are harmless in the case of guinea-pigs, others do not affect the health of mice. Cultures of the latter quality were continued during two years by Koch, Gaffky, and Löffler without any change or any return to virulence.

The behaviour of Bacilli which have been rapidly attenuated at 47–50° C. and at still higher temperatures is different; they recover their virulence when cultivated under the most favourable conditions.

It is true that a return from the attenuated to the virulent condition is not altogether excluded, even in forms which have been slowly attenuated. Pasteur affirms that if matter, which is not fatal to full-grown guinea-pigs but kills young individuals a day or two old, is taken from one of the latter and used to inoculate other guinea-pigs successively older, a degree of virulence is ultimately attained sufficient to kill old animals. Koch and his colleagues have not found these statements confirmed by their results, and though their experiments were somewhat differently arranged, yet they seem to show that no such fixed rule obtains in this matter as might be gathered from Pasteur's account. On the other hand the same observers have distinctly proved that there is a return to increased virulence in individual and not very similar cases.

Lastly, they have also proved that cases of the reverse kind occur, in which the virulence of a culture suddenly diminishes of its own accord, that is without any ascertained external cause; spores from material which killed rabbits and guinea-pigs produced eight weeks later a generation which did not injure these animals but was still fatal to mice. To this class of phenomena belongs perhaps an observation recently communicated by Prazmoswki, according to which a pure culture of Bacillus Anthracis in a nutrient solution entirely lost its virulence without apparent cause. I have myself seen the same or a similar case. Buchner's investigation, which bears on this point, will be noticed presently.

I have already observed that the form of the Bacillus remains unchanged in the virulent and attenuated states. In the main points this is always the case, though some modifications have been observed. Thus it is stated by Koch and his colleagues that the Bacillus which is only strong enough to kill mice fills the capillaries, especially of the lungs, in the form of long filaments, which may often be followed continuously from the capillaries into the larger microscopic vessels, while the more virulent Bacillus is usually present in the capillaries in the form of short rods.

In Prazmowski's observation the difference between the virulent and attenuated forms was, that the rods of the latter kind were motile during several generations, though their motion was slow and dragging in comparison, for example, with that of the hay-bacillus, and that they not only form flakes at the bottom of the nutrient solution which is clear above them, but also rise in it and make it turbid, and form on its surface 'thickish dirty-white films of a slimy consistence.' Exactly the same appearance has been observed by myself in meat-extract-solution; there the rods even up to the time of forming their spores were much less united into long filaments than in the virulent forms, and in the surface-films they lay in every direction, and were densely and irregularly compacted together into a felted mass. This mode of grouping is in appearance so unlike that of the common Bacillus, that we are naturally led to assume that we are dealing with a form very like Bacillus Anthracis, but yet specifically distinct from it, which has not allowed the latter to thrive in the solution, the Bacillus perhaps of Koch's malignant oedema; but the harmlessness of this form, even in the case of small rodents, is against this assumption.

Buchner undoubtedly observed the same phenomenon, when he grew Bacillus Anthracis through several generations in nutrient solutions composed of 1 per cent. of meat-extract, with or without the addition of sugar and peptone, and at a temperature of 35–37° C., and kept the cultures in constant movement by help of a rocking-apparatus in order to secure the largest possible supply of oxygen. The products of the cultivation gradually assumed the characters of Prazmowski's modified form.

Since this form in its want of virulence, its formation of superficial films, and its power of independent movement is more
like the hay-bacillus, B. subtilis, than the virulent form, Buchner maintained that the virulent B. Anthracis had been entirely
changed by cultivation into the harmless B. subtilis, and the
matter excited much attention, since there was here apparently
an evident case of the conversion of a species held to be distinct
into another.   But, as was shown on page 34, he has not yet
proved his point.   Buchner certainly tried the reverse process
also, and endeavoured to convert the harmless B. subtilis into
the virulent B. Anthracis by cultivating successive generations
of it in various solutions containing albumen, which I must not
enumerate here.   The results obtained were for the most part
distinctly negative, and the few which were said to be positive
are open to so many objections, that without insisting on strict
morphological proof we must consider them as quite uncertain.
Here too the morphological proof has been omitted.   It is quite
possible that the usually harmless B. subtilis may by breeding be
endowed with an exceptional virulence ; but its specific character
would no more be affected by this, than is that of B. Anthracis
by its attenuations, nor would the fact that the latter is the
ordinary exciting cause of anthrax be rendered at all doubtful.

Pasteur and Toussaint, led by experience derived from other
sources which will be noticed again presently, have attempted
with success to use the attenuated Bacillus of anthrax for protective inoculation against the virulent Bacillus.   If an animal is
inoculated with the Bacillus attenuated to the degree requisite
for it, that is for that species, it either does not sicken or it
sickens slightly and recovers from the disease.   It resists then
infection with less attenuated Bacillus, and at the next inoculation it resists the Bacillus which possesses the highest degree of
virulence.   The certainty of these results, and their special importance at the same time in reference to the practical art of
the breeder, is differently appreciated in different quarters;
Koch especially and his colleagues have brought forward well-

grounded objections to the praises bestowed on the school of Pasteur. We cannot enter further in this place into these practical questions. The fact, however, of the frequent success of protective inoculation is well established, and is abundantly confirmed even by those who are opposed to the exaggerated estimate of its practical importance. We record it therefore as a phenomenon of high scientific interest.

Having now become acquainted with these phenomena relative to Bacillus Anthracis and the disease which it produces, let us enquire how the injurious effects of the virulent Bacillus are brought about, and how the attenuation of the virulence and the operation of protective inoculation just described are to be explained.

In the present state of our knowledge we shall succeed in answering these questions best if we begin with the second. To prevent misunderstanding I will say beforehand most distinctly, that we can at present only make an attempt to answer these and all other questions, and must wait for our answers to be confirmed or amended by future investigation.

We begin therefore with the question of the explanation of protective inoculation, and we may formulate it a little differently and extend it by asking how it is that an animal is or becomes unsusceptible to, safe from the attacks of the injurious parasite. Metschnikoff has lately published some investigations which, if confirmed, bring us a step nearer to the understanding of this phenomenon. I report them because they seem to be trustworthy; I have not been able to repeat them myself.

We know that the blood of vertebrate animals contains red blood-corpuscles suspended in the fluid blood-plasma, and besides these colourless or white blood-corpuscles or blood-cells in considerably smaller quantity. The lower animals have no red blood-corpuscles, only the colourless blood-cells, which are uncoloured nucleated protoplasmic bodies. They possess a variety of remarkable qualities, but at present we are chiefly concerned with the fact that, like many other protoplasmic

bodies of similar structure, they are subject during their life to
constant changes of shape in their soft viscid substance, exhibiting undulatory movements of their outline and alternate
protrusion and retraction of processes (see Fig. 19). These
amoeboid movements, as they are called, are combined with the
power of taking up and absorbing solid bodies, or oil-drops, or
similar objects into their soft substance. If the foreign body
comes into contact with the surface of the amoeboid cell, the
latter puts out processes which embrace it, and gradually close
over it as the waves close over a drowning animal, so that it
lies at last inside the soft cell-substance. It may be cast out
again at some future time, but it may also suffer decomposition
inside the cell, be killed, and disappear.

In connection with these well-known facts, and also with the
observation made by him in the case of a disease in some small
crustaceans caused by the invasion of a peculiar Sprouting
Fungus, that the cells of the Fungus were absorbed by the colourless blood-cells of the animal and decomposed in it, that there
was a struggle, so to say, between the parasitic Fungus and
the amoeboid cells of the animal, Metschnikoff investigated
the behaviour of the colourless blood-cells of the vertebrate
animals to the Bacillus of anthrax. He found that the virulent
rods when introduced by inoculation into an animal liable to
take the fever, such as a rodent, were absorbed by the blood-
cells only in exceptional instances. They were readily absorbed
by the blood-cells of animals not liable to the disease, as frogs
and lizards, when the temperature was not artificially raised
(Fig. 19), and then disappeared inside the cells. The same thing
happened when susceptible animals were inoculated with Bacillus
Anthracis which had been attenuated to the harmless state.
Chauveau had already stated that the attenuated Bacilli pass into
the lungs and liver of the animal and there disappear. From all
these data we must assume with Metschnikoff that the Bacillus is
harmless because it is absorbed and destroyed by the blood-cells,
and injurious because this does not happen; or at least that it

bec omes harmless if the destruction by the blood-cells takes place more rapidly and to a greater extent than the growth and multiplication of the Bacillus, the converse being also true.

If these ideas are correct, and a normally virulent Bacillus is harmless after protective inoculation and in the absence of this is still virulent, we are driven to the further assumption that the protective inoculation produces this result by giving the blood-cells the power which they did not before possess of absorbing and

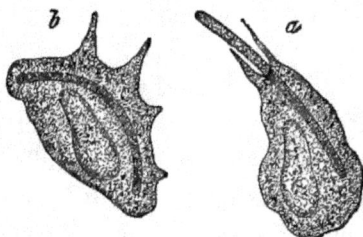

Fig. 19.

destroying virulent Bacilli. We have no distinct investigations into this point for our guidance, but if we once more accept the views set forth above we must almost necessarily assume that by the actual absorption of less virulent Bacilli the blood-cells of an animal successively acquire the power of absorbing and destroying the more virulent, which would not have been taken up without this preparation.

The security of an animal from infection by a dangerous parasite which has found its way into its blood, or the susceptibility to it, would accordingly depend on the reaction of the blood-cells on the parasite; and it would be possible to alter their power of reaction by gradually accustoming them as it were to a succession of more and more virulent individuals. In this way we obtain a partial explanation of the effects of protective inoculation with material of increasing virulence.

But now there is the further question, why are virulent Bacilli scarcely ever taken up by the blood-cells of an unprepared animal, while the attenuated ones are readily absorbed? Since

Fig. 19. *a* blood-cell of a frog in the act of engulphing a rod of Bacillus Anthracis, observed in the living state in a drop of aqueous humour. *b* the same a few minutes later; the shape of the cell is changed and the bacillus is completely engulphed. After Metschnikoff. Highly magnified.

we find no morphological or anatomical differences in the concurring parts in cases which differ furthest from one another, we can only assume that the cause of the difference in behaviour lies in material differences, differences in the chemical behaviour of the two objects. And since we are dealing on the one hand with portions of an animal which, as far as we can perceive, is not essentially altered in its collective properties, and on the other hand with a Bacillus which has its properties which are in this case the subject of direct observation essentially altered by attenuation, it follows that the changes in the chemical qualities must chiefly be on the side of the Bacillus. Nor is this at all inconsistent with the phenomena of protective inoculation, the accustoming the blood-cells, as was said, to the absorption of a succession of Bacilli each more virulent than the preceding one. On the contrary we know that other amoeboid protoplasmic bodies which absorb solid substances, for example the plasmodia of the Myxomycetes, do become habituated to contact with and probably also to the absorption of bodies with certain chemical qualities, though at first they hastily withdraw from contact with them; and without further arguments there is good reason for attributing the same power to the blood-cells, because they agree with plasmodia in all other qualities which have any bearing on this point.

No precise account can at present be given of the nature of the chemical differences between virulent and attenuated Bacilli, and what can be said about it will be mentioned presently. The proximate cause of the attenuation by Pasteur's method is to be sought not in the effect of oxygen but in the heightened temperature; this has been shown convincingly by Chauveau and by Koch and his colleagues, who call attention to the fact that the degree and permanence of the attenuation and also the time required for attaining it are directly dependent, other conditions being the same, on the temperature and even on small variations of temperature. We know nothing at present respecting the causes of the attenuations obtained by other methods than that of Pasteur and of the possible return of virulence.

If we proceed in the next place to enquire how the parasite causes disease, we find that it is not possible to give a decisive answer; still there are some definite facts and analogies leading to a conception of the matter which approaches very near to probability. The first fact is the appearance of the carbuncle of anthrax in human subjects infected by inoculation or through a wound. At the spot where the infection has taken place there appears at first a violent local inflammation of the skin, and it is not till 2–3 days later that the general symptoms supervene. The inflammation is specifically different from other violent inflammations of the skin, just as the local symptoms produced by a definite poison with peculiar effects differ from others which are caused by some other poison or by other causes. This, it appears to me, excludes the view sometimes expressed, that the Bacillus becomes the exciting cause of disease by giving rise to merely mechanical disturbances, or simply by withdrawing the oxygen from the living blood in which it vegetates; it is much more probable that the effect of the Bacillus is peculiar to itself, the result of a specific poison. If this is conceded, then the poison must issue, be excreted from the Bacillus, for it could have no effect if it remained in it. This agrees with Metschnikoff's observation that the same blood-cells readily absorb the Bacillus if it is not virulent, while if it is virulent it is virtually not absorbed. There must be something in the virulent Bacilli which there is not in the others, and it is probable, as was shown above, that this some-thing possesses distinct chemical properties, and it must be on the outside of or at least at the surface of the body of the Bacillus, for if it were only in the inside there would be no reaction with the blood-cells upon their coming into contact with the Bacillus.

We know nothing of the real nature of the poison which is thus supposed to be excreted by the Bacillus. Attempts to

Missing Page

Missing Page

bottom of Pasteur's culture-fluid when vegetation has ceased
and the nutrient-material is exhausted, and will remain there
alive and capable of renewed vegetation in a suitable substratum
for about eight months if the air has access to it, for a longer time
if air is excluded, as in hermetically sealed flasks.

If the Micrococcus is taken fresh from a sick or dead bird
or from its excrements, or from an artificial culture absolutely free from any diseased part of the bird, and introduced
into a healthy bird, the minutest possible quantity of it, exhibiting ordinary growth and multiplication, will reproduce the
disease. Infection is obtained by inoculation in or beneath
the skin, and by introduction of the Micrococcus with the
food into the digestive canal. Pasteur succeeded in com-
municating the disease by inoculation to mammals as well as
to birds; rabbits thus inoculated died; in the case of guinea-
pigs abscesses only formed at the place of inoculation, and
these contained a large quantity of the Micrococcus, but
they did not spread and ultimately healed. Kitt has extended
these experiments with similar results to mice and sheep and to
a horse.

Enough has been said to show that in this Bacillus, as in that
of anthrax, we have a facultative parasite specifically capable of
causing disease; but its history and manner of life, especially
as regards the saprophytic sections of its development are not as
well known to us as those of Bacillus Anthracis.

Pasteur further discovered that the Micrococcus loses to some
extent its power of infection when it is kept some time exposed
to the air; the number of successful inoculations and the se-
verity of the attack caused by them diminish with the age of the
matter used in inoculation. The cases mentioned before of
slighter attacks of the disease ending in recovery are chiefly
cases of inoculation of this kind. We may say therefore in
words which have been used before that age produces an
attenuation of the power of infection or virulence of the
Micrococcus.

Individual birds, which had recovered from the disease, were generally but not always found to be no longer susceptible to virulent infection, to be secure against a fresh attack, and on this Pasteur founded the ideas with regard to protective inoculation and his method of employing it which he proceeded to extend to anthrax also.

When the fluid of a fresh culture in broth was separated by filtration from the Micrococcus, which cannot be done by filtration through paper, for the Bacteria invariably pass with it through the paper, but can be managed by means of filtration through porous earthenware, the fluid was not in a condition to produce the disease in the perfect form, not even when all the constituents contained in solution in 120 grammes of it were injected into the blood of a fowl. But one characteristic symptom of the disease, the stupor, was produced; the birds were sleepy and as if stupefied after infection, continuing in this state about four hours, and then returning to their normal state of health.

The observation, if established, shows that in this case a narcotic poison separable from the Bacterium was actually disengaged from it, and this is the reason why the fowl-cholera is especially instructive in judging of the effects of parasites of this kind in the production of disease. That the effect of the poison in these experiments was comparatively slight and transitory, is to be explained by the small amount of it in the fluid, and by the fact that like other poisons it is either decomposed in the infected fowl or is withdrawn from it with the normal secretions. The case is different when the poisonous organism itself is present in the fowl, apart from the circumstance that the conditions are then probably more favourable for the formation of the poison. While the poison is being perhaps constantly decomposed within the fowl or is being removed with the normal secretions, it is constantly being produced by the parasite, and what has been removed is replaced; thus the symptoms of the disease necessarily become more permanent and more severe, and ultimately also more complicated. Further complications also arising from

the more strictly mechanical effects of the parasite are of course
not impossible.

XIII.

**Causal connection of parasitic Bacteria with infectious
diseases, especially in warm-blooded animals. Introduction. Relapsing fever. Tuberculosis. Gonorrhoea.
Cholera. Traumatic infectious diseases. Erysipelas.
Trachoma. Pneumonia. Leprosy. Syphilis. Cattle-
plague. Malaria. Typhoid fever. Diphtheria. Infectious diseases in which the presence of contagium vivum has not been demonstrated.**

1. I SHOULD have wished to have added to the foregoing
two examples of facultative parasites which cause disease, one
example at least of a strictly obligate parasite, but I am unable
to find even one which is sufficiently well known to allow a
detailed account to be given of it. All that can be produced on
this point will therefore be included in the following summary,
which is intended to contain in a few words the most important
part of the knowledge which we at present possess of the nature
of Bacteria as the exciting causes of infectious diseases in warm-
blooded animals, and especially in man (58).

By the term infectious diseases we mean all those forms of
disease which are only found where they are conveyed from a
sick person to a healthy one so far as the particular disease is
concerned, or the origin of which is confined to localities of a
particular character. The former kind are known as contagious
diseases, such as scarlet fever, measles, small-pox; the latter, of
which malarial fever is the best known example, are termed
miasmatic diseases. If the two conditions are combined, we may
speak of miasmo-contagious disease and that in two senses; we
may mean firstly that a disease may be caught in certain localities,
or by infection from person to person independently of locality
of miasmatic character, or secondly that a disease is indeed

contagious, but requires the existence of previous miasmatic infection in the person who is to take it. It should be observed also, that up to recent times the term infectious disease was only used when the exciting cause of infection, the contagium or miasma, was only imperfectly understood. If disease was caused by known parasites, such as lice or entozoa, transferable from one person to another or only to be procured in certain localities with special characters, it was not called infectious but parasitic.

It was natural that some ideas should be entertained respecting the general qualities of the unknown and invisible contagia and miasmata, and it was assumed, not without reason, that they were special particles of matter which were efficacious for infection in a state of the most minute subdivision and in the most infinitesimal quantities.

The qualities of living beings were for a long time ascribed by some observers to these infectious particles or contagia, as we may now usually call them, and first of all in the period of history from which we have received the terms 'contagium vivum' or 'animatum,' then used in a somewhat obscure and indefinite sense. The traditional expression contagium vivum received a more precise meaning in 1840 from Henle, who in his 'Pathologischen Untersuchungen' showed clearly and distinctly that the contagia till then invisible must be regarded as living organisms, and gave his reasons for this view. His argument may be briefly stated in the following manner. The contagia have the power, possessed as far as we know by living creatures only, of growing under favourable conditions, and of multiplying at the expense of some other substance than their own and therefore of assimilating that substance. The quantity, certainly minute, of the contagium which communicates infection to any one who pays a hasty visit to a person suffering from an infectious disease, has the power of multiplying enormously in the body of the infected person, for the latter is able to give the infection to an unlimited number of healthy but susceptible persons, and therefore to part again for an unlimited number of times with the same minute portion of contagium which he himself received. But if we are forced to recognise the characteristic qualities of living beings in these contagia, there is no good reason why we should not regard them as real living beings, parasites. For the only general distinction between their mode of appearance and operation and that of parasites is, that the parasites with which we are acquainted have been seen and the contagia have not. That this may be due to imperfect observation is shown by the experiments on the itch in 1840, in which the contagium, the itch-mite, though almost visible without magnifying power, was long at least misunderstood. It was only a short time before that the microscopic Fungus, Achorion, which causes favus, was unexpectedly discovered, as well as the Fungus which gives rise to the infectious disease in the caterpillar of the silkworm known as muscardine. Other and similar cases occurred at a later time, and among them that of the discovery of the Trichinae between 1850 and 1860, a very remarkable instance of a contagious parasite long overlooked. Henle repeated his statements in 1853 in his 'Rationelle Pathologie,' but for reasons which it is not our business to examine here, his views on animal pathology met with little attention or approval.

It was in connection with plant-pathology that Henle's views were first destined to further development, and obtained a firmer footing. It is true that the botanists who occupied themselves with the diseases of plants knew nothing of Henle's pathological writings, but made independent efforts to carry on some first attempts which had been made with distinguished success in the beginning of the century. But they did in fact strike upon the path indicated by Henle, and the constant advance made after, about the year 1850, resulted not only in the tracing back of all infectious diseases in plants to parasites as their exciting cause, but in proving that most of the diseases of plants are due to parasitic infection. It may now certainly be admitted that the task was comparatively easy in the vegetable kingdom, partly because the structure of plants makes them more accessible to research, partly because most of the parasites which infect them are true Fungi, and considerably larger than most of the contagia of animal bodies.

From this time observers in the domain of animal pathology, partly influenced, more or less, by these discoveries in botany, and partly in consequence of the revival of the vitalistic theory of fermentation by Pasteur about the year 1860, returned to Henle's vitalistic theory of contagion. Henle himself, in the exposition of his views, had already indicated the points of comparison between his own theory and the theory of fermentation founded at that time by Cagniard-Latour and Schwann.

Under the influence, as he expressly says, of Pasteur's writings, Davaine recalled to mind the little rods first seen by his teacher, Rayer, in the blood of an animal suffering from anthrax, and actually discovered in them the exciting cause of the disease, which may be taken as a type of an infectious disease both contagious and miasmatic also, in so far as it originates, as has been said, in anthrax-districts. This was, in 1863, a very important confirmation of Henle's theory, inasmuch as a very small parasite, not very easy of observation at that time, was recognised as a contagium. It was some time before much further advance was made. Rather the too great zeal of inexperienced observers, especially excited by the cholera-epidemic of 1866, led to a so-called searching for parasites, barren of all but mischievous results, which was the more calculated to repel more earnest observers because for a time it attracted some measure of applause. These are things which are now long passed away and require no further notice.

Since about the year 1870 more general attention has been again directed to these questions. The number of publications discussing or bearing upon them is rapidly increasing, and we cannot follow them here in detail. Prominent among them, as new and especially suggestive attempts to deal with the subject, are Cohn's and Billroth's works already mentioned (1, 6), and those of von Recklinghausen and Klebs on the more specially pathological side; it is the merit of Klebs more particularly to have clearly indicated not only the nature of the problem as conceived by Henle, whom he expressly follows, but also the details of the ways and methods to be adopted for its solution, and to have pursued them himself, though sometimes perhaps with more zeal than discretion. Pasteur and his school pursued the same course independently. Thus questions and experiments were framed and knowledge acquired with constantly increasing precision and results. The latest advance to be recorded begins with the participation of Robert Koch in the work of research since 1876. He may claim to have gone forward as a highly intelligent observer on the paths marked out by his predecessors without precipitancy, and with careful use of all improvements in morphological investigation and in microscopical and experimental methods. Hence he was the first to obtain clear results in cases which up to that time had always been disputed, as is shown by our account in a former lecture of the aetiology of anthrax, the final settlement of which is due to his investigations; he has also shown what must be done to ensure progress in researches of this kind.

The result of all these efforts is the same as that which was arrived at in plant-pathology thirty years before. Firstly, it is certain that in a number of cases the contagium is a microscopic parasite, and that certain diseases, the infectious nature of which was formerly denied or was doubtful, are infectious diseases of this kind. Secondly, the same conclusion is rendered at least highly probable in other cases. Thirdly, there remains a very considerable number of diseases in which the parasite has been sought for, but has either not yet been found, or its existence is doubtful.

Further, it has been proved that, with some exceptions, especially of skin-diseases and similar affections in which true Fungi of a relatively large size are concerned, by far the most important parasites of contagion in warm - blooded animals hitherto certainly determined are Bacteria.

We may note as consequences of what has now been said, firstly, that Henle's doctrine has become a widely-accepted dogma. There is no objection to this, if we put in the place of belief that intelligent personal conviction which is distinctly directed toward a particular view, but does not lose sight of the possibility that it may some day be corrected or altered. That the parasite required by the theory has not yet been found is no reason for abandoning it, for the parasite may very easily have been overlooked, owing to its extreme minuteness, or its power of refraction, or because the observer has not learnt the right place or time to look for it. When Henle founded his doctrine in 1840, the Bacillus of anthrax had never been seen; the Trichinae had been seen, but no one suspected that they were the cause of disease.

The second result is that at present in almost every doubtful or questionable case, it is only for Bacteria that search is made. This is wrong in principle; it may be practically right to search for such forms as present experience shows are the most likely to be found. But it should be remembered that organisms of another kind may make their appearance unexpectedly, about which we at present perhaps know very little. It is not so long ago, that we knew very little about Bacteria and expected them as little. That this is not idle talk is shown by some surprising experiences recorded in connection with plant-pathology and by the history of pébrine which will be noticed presently.

Thirdly, if belief is stronger than the critical faculty, there is great danger of concluding at once from the presence of a Bacterium that it is the exciting cause of disease for which we may be seeking. From what we learnt in Lecture V on the wide diffusion of Bacteria possessing full powers of development it will at once be seen that Bacteria may be developed in a diseased body before or after death, and that a particular form may be present as a characteristic feature and even constantly and exclusively in a particular disease, and therefore have high diagnostic value, without being the contagium which causes the disease. To make sure of this it is absolutely necessary to experiment with pure material and to obtain a clear positive result; there must therefore be a pure separation of the parasite to be examined from all admixtures, a pure infection of the proper subject for experiment with pure matter, and the strictest control and criticism of the result. The example of anthrax described above will illustrate these rules. Without successful experimentation there is always a gap in the proof which cannot be filled up by other arguments, however well adapted they may be to serve as the foundation of a personal conviction. It is true that the latter may persist in spite of defective experimental proof. A parasite, as has been before explained, does not thrive, or does not thrive equally well in every host-species; it can attack and cause disease in one and not in another. The experiment therefore in the case in question may give no positive result, because the right, that is the susceptible, species of warm-blooded animal was not employed. This point must be specially attended to in the case of infectious diseases which attack human beings chiefly. We cannot or must not experiment freely with human beings, but must trust in the main to experiment on other warm-blooded animals, and this may be the only reason why that in certain cases, some of which will be noticed further on, the results of experiment have as yet remained doubtful or negative.

Enough has been said to give inexperienced persons an idea beforehand of the reasons why we may have to speak here of doubtful cases and doubtful statements.

We will now proceed to a consideration of the facts. Our object, as was said before, is to bring forward whatever is most important in connection with the Bacteria which are the exciting causes of disease. A minute discussion of the diseases themselves does not fall within the plan of these lectures, and must be sought in medical publications.

Whatever peculiarity there may be in each individual case, we are constantly confronted in dealing with the main points of our subject with similar facts and questions to those which were considered at some length in connection with anthrax and fowl-cholera ; they form part of the great series of facts and questions relating to parasitism, of which it was attempted to give a concise review in Lecture X.

With this brief reference to these previous expositions we now proceed to consider a few comparatively well-known cases.

2. Relapsing fever, febris recurrens (59), is a disease which is widely spread in Asia and Africa, is endemic in Russian Poland and Ireland, and sometimes finds its way into other countries of Europe. It is communicated directly from the body of one person to another or through the intervention of articles of daily use. In 5–7 days after infection violent fever sets in with other symptoms which need not be described here, and usually lasts another 5–7 days, and is then followed by a period of absence of fever for about the same number of days. Then comes a relapse into the fever state, and the same alternation may be repeated several times, usually with a favourable ending.

During the attack a slender Spirillum, resembling Spirochaete Cohnii (Fig. 16, *e*), sometimes 40 μ in length and exhibiting active movements, is found in abundance in the blood of the patient which is often of a dark-red colour; discovered by Obermeier in 1873 it was named after him Spirochaete Obermeieri; it disappears during the interval when there is no fever.

The disease is conveyed to men and monkeys when they are inoculated with the blood of a fever-patient containing Spirochaete. Blood taken during the interval of freedom from fever and therefore free from the Spirochaete does not produce the disorder after inoculation. Experiments in the inoculation of other animals were always without result. Attempts to cultivate the Spirochaete outside the body of the animal have not as yet been successful.

From these facts it may properly be assumed that the Spirochaete is the contagium of relapsing fever, though we are still

very imperfectly acquainted with its life-history, for we have
no certain information as to its place of sojourn during the intervals of the fever, the form and mode of its conveyance from
one person to another, the formation of spores if any, or other
resting states.

3. One of the most important results of the researches into
the Bacteria which are the exciting causes of disease is the discovery of the contagium of tuberculosis, the Bacillus of tubercle
long since rendered familiar to us by Koch's publications (60).
Tuberculosis has received its name from the formation of new
substance or the degeneration by which it is characterised, and
which appear in the form of small knots or tubercles in the
tissue of the organs.   The formation of tubercle is best known
in the lungs as pulmonary-tuberculosis, pulmonary consumption,
but it may occur in any organ and the lymphatic glands are
particularly subject to it.

Tuberculosis attacks warm-blooded animals of every species
as well as man, and especially our ordinary domestic animals
and those used for experiments.   The tuberculosis of horned
cattle is known by the name of pearly disease, bovine tuberculosis.
Different species show different degrees of susceptibility; the
field-mouse is highly susceptible to infection, the domestic mouse
only slightly. The primary anatomical changes in the formation
of tubercle are in all cases the same.   The succeeding ones and
the general character of the disease may be very dissimilar.

In tubercle, at least in the fresh state, Koch and simultaneously with him Baumgarten demonstrated the presence of a
characteristic rod-shaped Bacillus. According to these observers
it is always there, though in very unequal quantities in different
cases.   It passes into the ejected matter, the sputum of pulmonary tuberculosis, and may be found in it. With due care it may
be kept pure and be cultivated pure through repeated generations
on stiffened blood-serum or in infusion of meat.

Tubercular matter containing Bacilli, or still better the purely
cultivated Bacillus, introduced beneath the skin of susceptible

animals or injected into a blood-vessel or into a cavity of the body, or pure Bacillus-material inhaled in a state of fine division suspended in water, resulted in the formation of tubercle with its consequences in every case without exception — Koch experimented on 217 individuals of susceptible species of animals (rabbits, guinea-pigs, cats, field-mice), besides animals used in control-experiments and individuals of less susceptible species—and the Bacillus was in every case found in the tubercles. In every case also the place where the tubercle appeared, its frequency and distribution, and the line of distribution through the body answered the expectations founded on the mode of infection and the spot where the infecting matter was applied. These results, still further strengthened by control-experiments, established the infectious nature of tuberculosis, and the contagious character of the Bacillus, which had indeed been previously concluded on other grounds.

Such observations on the Bacillus itself as have been published leave much to be desired as regards its morphology. Observers have been generally satisfied with proving its presence, and the proof has been rendered easy by its peculiar behaviour with aniline colouring matters. In contrast to the great majority of other known Bacteria, it slowly and with difficulty takes in alkaline solution of methylene blue or saturated solution of methyl-violet, absorbing them only in the space of several hours or after being heated, but it obstinately retains the colour it has acquired, while other Bacteria are quickly decolorised by certain reagents, for example, dilute nitric acid. By this peculiarity in their behaviour as well as by their shape and size, the Bacilli are comparatively easy to recognise and distinguish from other species. In form they are slender rods which are sometimes curved or bent at an angle, and reach a length of $1 \cdot 5$–$3 \cdot 5 \mu$. Neither in the natural nor in the coloured state can transverse septation as a rule be observed. Endogenetic spores are found in them, both in cultures and in the body and sputa of diseased animals; these according to Koch's brief account must answer

to the spores of endosporous Bacilli, but no further description
is given of them. Judging from the figures of sporogenous specimens, and assuming that this species does not differ altogether
in behaviour from other endosporous Bacteria (see page 15),
the rods must be segmented in the same way as those of Bacillus
Megaterium described above, for they are represented with 4–6
spores standing close in a row one above another, as in Fig. 1, *r*.
If our assumption and the account given is correct, each of these
spores must lie in a short segment-cell.   This agrees with the
fact, that in coloured preparations the rods are sometimes found
divided by narrow hyaline transverse septa into a series of segments not longer than broad, such as Zopf figures in his third
edition. To call these segments Cocci is only playing with words.
If young vegetating rods are divided into long segments, this is
another point of agreement with Bacillus Megaterium.   After
this description I abstain from giving a figure of this species; a
drawing of what has been at present seen would represent a
simple or interrupted black stroke.   In Fig. 1, *b–f* and *r* correspond with our present knowledge of the form of the Bacillus of
tubercle, except that the length of the rods in the latter species
is on an average not greater than the breadth of those of Bacillus
Megaterium in the figure.

The living rods, according to Koch, are not motile.   When
grown on stiffened blood-serum they do not liquefy it, but
remain on the surface, and there even when developed in comparative abundance they form thin dry scales of small extension,
which are shown under the microscope to consist of sinuously
curved swarms and bundles of single rods.

The Bacillus of tubercle grows slowly as compared with most
other Bacteria, and in this respect resembles the Bacterium of
kefir.   In cultures on serum 10–15 days elapse before growth
can be detected by the unaided eye.   The result of an inoculation is not apparent in less than 2–8 weeks.

The attempts to cultivate this species outside the living animal
on any other nutrient substrata than those above-mentioned

have not been successful; the cardinal points for the temperatures of vegetation are those given on page 51.

The Bacillus of tubercle offers a somewhat high degree of
resistance to injurious influences from without, and is thus able
to preserve its powers of infection. It can bear temperatures
approaching the boiling point, though it is soon killed if it is
heated in a thoroughly moist condition. It was not affected by
desiccation during a period of 186 days, or by being kept in
putrefying sputum for 43 days. The experiments on its powers
of resistance have chiefly been made with sputum containing the
Bacilli. No attempt has been made to determine precisely how
far these powers of endurance are confined to the spores or
belong also to the vegetative rods, but our experience in other
cases would lead us to suppose that they belong chiefly to the
spores.

These facts taken together give a satisfactory explanation of
the appearance of tuberculosis as the result of infection with the
Bacillus. Every one knows how widely spread the disease is,
even if we think only of pulmonary tuberculosis; a seventh part
on an average of the deaths of human beings are caused by this
form of the disease. The Bacillus is generally present, capable
of development and in a virulent condition, in the excretions of
those suffering from tuberculosis. The expectorations of consumptive persons often during months and years must be especially but by no means exclusively taken into account in this
connection. The Bacillus was absent from 44 only of the 982
specimens of sputa examined by Gaffky. It is clear that the
Bacillus is communicated in large numbers to these excretions,
and that when they dry up, it must be disseminated with the
dust and in other ways. There is therefore abundant opportunity for infection in the ordinary intercourse of human beings.
We need not enter more minutely into this point, and to discuss
the mode in which the Bacillus spreads in the body which has
once become infected would also lead us too far into medical
details. That a good deal depends on the susceptibility of the

subject to be infected to the results of infection is shown by the
fact, that in sick-rooms and institutions in which consumptive
patients remain the year through, it is not every one who is suc-
cessfully infected with tuberculosis.   This fact is in accordance
with the general knowledge which we possess of individual or
specific differences in susceptibility to the attacks of parasites.

The foregoing facts and views are not affected by the statements of Malassez and Vignal, who describe a form of tuberculosis with a copious growth of Micrococcus, which they call
'tuberculose zoogloique,' and in which they sometimes found
the Bacillus and sometimes not.   Supposing these accounts to
be correct, we must assume with the clear results of Koch's investigations before us that there is either some complication
here, or that we are dealing with some disease resembling
ordinary tuberculosis but different from it in so far as it is caused
by a different parasite.

4. Gonorrhoeal affections (61).   These are inflammations
of the mucous membrane of the urethra and of the conjunctiva of the eye accompanied by suppuration and occurring
in the human species.   Purulent conjunctivitis in new-born
children, ophthalmia neonatorum, may certainly be classed
with them.

One of the characteristic peculiarities of these maladies is that
they are highly infectious, and it has long been known that infection is due to the purulent secretion of the patient.   The
infection of healthy human eyes takes place, as Hirschberg says,
with the certainty of a physical experiment.   With the same
certainty there is found in the infectious matter a large Micrococcus, which was discovered by Neisser and named by him
Gonococcus (Fig. 20), chiefly appearing to be attached to the
surface of the epithelial and pus-cells, according to recent
observations really penetrating a slight distance into the body
of the cells, less often lying between them.   It should be added
that the number of the cells beset with the Gonococcus is always
relatively small and varies from case to case.

The cells of the Gonococcus are roundish in shape and of some size, about 0·8 μ in diameter, often attached together in pairs corresponding to their partitions, separated in the full-grown state by a hyaline gelatinous intervening substance, and often distributed in large numbers and at tolerably regular distances on the surface of the pus-cells. It is uncertain whether this superficial arrangement is due to successive partitions taking place alternately in two directions, or to a corresponding displacement with the partition always in the same direction; I see nothing in the observed facts to compel us to adopt the first assumption.

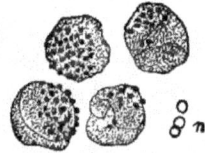

Fig. 20.

Gonococcus is not found in other inflammations of the mucous membranes in question, and other Bacteria do not give rise to the phenomena of gonorrhoea. These facts make it probable with the aid of analogy that the infectious nature of the gonorrhoeic secretion is due to the presence of the Coccus, that this is the active contagium.

Warm-blooded animals other than man are, as far we know, either not susceptible to gonorrhoeic infection or take it with difficulty; a very large majority of the experiments on animals with the secretion from the eye were unsuccessful.

Cultures of Micrococcus Gonococcus outside the living patient seldom succeed. Yet some are said to have been successful, those for example of Hausmann from the secretion from the eye of an infant on stiffened blood-serum, and those of Bockhardt and Bumm from the secretion from the urethra;

Fig. 20. Micrococcus Gonococcus, Neisser. From the secretion from the conjunctiva of a child treated for Ophthalmia neonatorum. Four pus-cells with Micrococcus attached, from a preparation coloured with methyl-violet. The pale-coloured pus-cells with their nuclei are faintly shown in the drawing in order to make the Micrococcus more apparent. *n* outline of an isolated cell of Micrococcus and of a pair of cells formed by bipartition. Magn. 600 times, with the exception of *n*, which is more highly magnified.

the careful observations of the latter author leave no room for doubt. The introduction of the Coccus from a pure culture into the eye of a new-born rabbit by Hausmann, and into the eye and urethra of a human subject by Bockhardt and Bumm, communicated the infection. It appears from Bumm's observations on the conjunctiva of the human eye, that the Micrococcus penetrates between the epithelial cells into the papillary body of the mucous membrane, multiplies and spreads in these spots and later also in the purulent secretion, and is ultimately stopped in its further advance and got rid of by regeneration of the epithelium and secretion of pus. The case examined by Bockhardt showed more complicated phenomena. Bumm's excellent monograph should be consulted for the details.

I have put together relapsing fever, tuberculosis, and gonorrhoea, different as they are from one another, because if we once put the doubts and the gaps in our knowledge on one side, and take probability for certainty, they supply us with examples of actual obligate parasitic Bacteria.

Spirochaete Obermeieri is, as far as we at present know, strictly obligate, inasmuch as it can only be conveyed from one person to another without digression into saprophytism, and is confined to men and apes.

The Bacillus of tubercle and Gonococcus may certainly be cultivated as saprophytes, a facultative saprophytism cannot be denied them. But this character can scarcely be taken into consideration in their case ; not in that of the Bacillus of tubercle, as Koch urges, because the conditions of its vegetation as a saprophyte are of such a kind and so limited, that they will scarcely ever be found except in an apparatus contrived for the special purpose, nor yet in that of Gonococcus for similar reasons. This follows at once from our experience in general, and it follows further that the resisting power of the Gonococcus is very small, and its dissemination for the purpose of infection, as for example in dust after desiccation, is not worth consideration. Gonorrhoeic affections are scarcely less common than tuberculosis ; their secretions are scattered about and the Gonococcus with them. If the Gonococcus were capable of saprophytic vegetation under ordinary natural conditions, it is hardly to be supposed that infection would not sometimes take place in some other way than from one person to another. This, however, notwithstanding some quite doubtful and isolated accounts, is not the case.

5. Asiatic cholera (75) may now be very fairly reckoned among the comparatively well-known infectious diseases, which we are at present considering. As early as the beginning of the year 1850, Pacini believed that he had found a contagium vivum of this disorder in the Bacteria, or Vibrios as he terms them, which he observed in the intestinal canal and its evacuations. Some time after (1867) Klob examined the contents of the intestinal canal, and the evacuations of patients suffering from Asiatic cholera, and likewise found Bacteria present in them in considerable quantities ; and starting with the assumption that these organisms are instruments of decomposition, he showed it to be probable that these Bacteria set up the disease in the intestinal canal, and spread it from thence to other parts of the body. Our knowledge of the Bacteria had not then reached the point at which attempts could be made to distinguish more precisely between the various forms found in the intestine and in the dejecta and to separate them from one another. The absurd notion, entertained in other quarters between the years 1860 and 1870, of referring the cholera-contagium, Bacteria included, to ordinary moulds and hypothetical parasites of the rice-plant, and the fact that Bacteria apparently quite similar to those of Klob were found in the intestinal canal of persons who were not suffering from cholera, had the effect of withdrawing attention once more from these and other efforts to discover the contagium vivum of this disease. In India, the perennial home of the pestilence, the researches conducted at a later time by English physicians gave no certain and positive results.

Our knowledge of the Bacteria and of the real existence of contagia viva was in a much more advanced state, when the outbreak of the epidemic in Egypt in 1883 led to a fresh resumption of the question. R. Koch, the most experienced investigator of the subject, pursued his enquiries in Egypt and in India, and there became acquainted with a distinctly marked Bacterium-form which is found in the intestinal canal in fresh cholera-cases, and was once observed also in a water-tank in a cholera-district. This Bacterium he suspected to be the specific contagium or miasma of the Indian pestilence; we will for the present call it a Spirillum.

According to the facts as at present known, there can scarcely be a doubt that Koch's Spirillum is really the contagium vivum of Asiatic cholera. First of all, its constant presence—for we may call it constant—in the small intestine or in the evacuations of cholera-patients is acknowledged by all observers, even by those who do not accept Koch's views. It is sometimes found almost as in the state of a pure culture in the mucus of the intestine in bodies examined immediately after death; under other circumstances certainly it is less pure and abundant. In the exceptional cases in which it was not found, either strict search was confessedly not made, or it may have been overlooked or have actually disappeared, especially at an advanced stage of the disease, and therefore have once been present. Koch's Spirillum has never been found either in the intestinal canal or in any other part of the body in any disease except Asiatic cholera.

The Spirillum of cholera can readily be cultivated pure as a saprophyte, as will be noticed again further on. Attempts to inoculate animals with pure living material of this kind gave at first only negative or in the most favourable case uncertain results, and this is especially true with regard to the experiments in which it was sought to convey infection with the food. It was found that the Spirilla were killed by the acid gastric juice, or were rendered inoperative by some other causes. But a change in the mode of conducting the experiment led to a positive result.

Nicati and Rietsch and van Ermengem avoided the passage of the stomach and introduced the Spirillum by injection directly into the small intestine. In van Ermengem's experiments the Spirillum, which had been cultivated in meat-broth or serum, was injected into the duodenum of some guinea-pigs, eleven in number, in small quantities,—a single drop or a much smaller portion of the fluid. Of the eleven, one died soon after the operation, nine in two to six days after infection; the eleventh, which had received 'about one-fiftieth of a drop,' recovered after a short illness.

The phenomena of the disease and the state of things as shown by dissection corresponded, according to van Ermengem's account, in all essential particulars with those of Asiatic cholera, making the necessary allowance for the difference between the human being and a guinea-pig. The Spirillum vegetated abundantly in every case in the intestine of the infected animal, either pure or mixed with other Bacteria. A drop of fluid containing the Spirillum from the intestine of the animal, communicated the same form of disease to sound animals when injected into their duodenum. Lastly, when fluids containing other Bacteria were injected into the duodenum as a test-experiment, no cholera-symptoms appeared, and usually no disturbance of the ordinary health.

I have here put these experiments in the front place, because they could be most simply and briefly described. Other observers, Koch especially and Doyen, obtained the same positive result by introducing the Spirilla with food after the acidity of the contents of the stomach had been neutralised by an alkaline fluid, and further by increasing the predisposition of the animals for the infection by administering opium and alcohol, in accordance with an observation of Koch's. I must limit myself here to these few remarks in proof of the success of the attempts to communicate the infection, and refer to the special literature for further details.

It appears from the accounts which we possess that the

M

Spirillum of cholera vegetates invariably in the intestine of the patient, both in the mucus of the intestine and also, according to some observers, penetrating into the tissue of the mucous membrane. Neither it nor any other Bacteria are found in other organs of those who have died of the disease, according to Koch and most other observers. But Doyen attests its presence in kidneys and liver, and van Ermengem found it in the blood-channel of three of the animals in his experiment before or immediately after death.

On the strength of the observation that the Spirillum occurs only in the intestinal canal, Koch's view is very generally thought to be highly probable, that it produces a very powerful poison there, and that it is this poison which being absorbed from the intestine causes the severe general symptoms of cholera. If it should be proved that the Spirillum is carried through the body with the flow of blood, this assumption must at least receive some modification, and in any case it still requires more distinct proof.

As regards its shape, Koch's cholera-contagium appears, when its segmentation is most perfect, in the form of spirally-twisted rods or filaments, closely resembling the Spirilla figured on page 82, and of very unequal length and number of spirals. The filament is about $0.5\,\mu$ in thickness, but it is not possible to give the exact measurement; the width of the turns of the spiral is about the same as the thickness of the filament, or less; the steepness of the individual turns varies. The filament is composed of segments or segment-cells, which are about as long as a half-turn of the spiral, and each of them, therefore, is a more or less curved rod. A separation of the segments from one another does actually and as a rule take place soon after every division, if the Spirillum is actively vegetating in a gelatinous nutrient substratum (gelatine, agar) or on the mucous membrane of the intestine; in these places, therefore, the Bacterium takes the form of crooked rods, which are either single or united together into short rows; the shape of these

rods was naturally compared by Koch to that of a comma, and
he therefore called them comma-rods, comma-bacilli. In good
nutrient solutions, such as meat-broth, and in old gelatine
cultures, the segments more frequently remain united together
into long unbroken and apparently unsegmented spirals. In
both forms the Spirillum has the power of movement, the single
rods being more active than the longer spiral filaments, especially
when they have grown in old gelatine-cultures.

Hueppe has observed in old cultures a further phenomenon,
which must be termed spore-formation,—the formation in fact
of arthrospores. The spiral filaments, beginning at intercalary
spots, divide for a certain distance into spherical segments,
which are a little thicker than the vegetating cells, are more
highly refringent, and are separated or held together by thinner
gelatinous envelopes. In this form they do not divide, but if
supplied with fresh nutriment they may at a later period
develope again into comma-rods, and by so doing they justify
their claim to be called spores. Spores of this kind appear to
have been seen, but not rightly understood, by former observers. They are quite distinct from the formations described
by Ferran. These are to be seen when spiral filaments in old
cultures swell up irregularly and shapelessly, or form round
bladders at their extremities, and then, as later observers have
unanimously declared, die off and disappear. They are connected, therefore, simply with the retrogressive or involution-
forms common among Bacteria, and noticed above on page 10.
Ferran's sensational descriptions of these objects are inconceivable to every sensible man with any pretension to a scientific
education. They have no other significance than that of a
warning example of the follies a man may commit, when he is
bent on making faulty observations seem important to himself
and others by the use of names and technical expressions which
he does not understand.

With respect to the biological characters of the Spirillum of
cholera, it is no longer needful from the accounts which we

have of it to refer expressly to its facultative saprophytism. Its saprophytic vegetation requires an abundant supply of oxygen. If cultivated with this upon a suitable moist substratum it developes rapidly and copiously, taking the place of any competitors it may encounter; but after some days the energy of its growth rapidly diminishes, perhaps in consequence of the disturbing influence of its own decomposition-products. These phenomena were most strikingly exhibited in cultures on moist linen, which was selected for practical reasons. The optimum temperature for vegetation is, as was stated above on page 50, that of the body of a warm-blooded animal, about $37°$ C., but $20-25°$ C. is sufficient for good development. Death ensues without fail in a fluid heated to $50-55°$ C. The Spirillum is not killed by being cooled to or below the freezing point, even if this temperature is maintained some hours. Perfect desiccation kills the vegetating Spirillum in less than twenty-four hours; the arthrosporous Spirilla on the contrary, according to Hueppe's direct observation, continue capable of germination for four weeks after desiccation. On the other hand, as it is a matter of observation that new vigorously vegetating generations may proceed from desiccated cultures after a longer period than this, and for almost ten months, Hueppe suspects, on good though not quite conclusive grounds, that the new growths always spring from arthrosporous Spirilla, and that these are the specifically resistent resting states of the Bacterium of cholera. I have already said all that is here necessary concerning the food-requirements of the Spirillum, and on the unfavourable or even fatal effect upon it of the acid reaction of the substratum.

The phenomena in the life of the Spirillum, which have now been described, supply the needful explanation of the chief facts of experience in connection with cholera as an infectious disease, especially its claim to be indigenous in India, its introduction into other countries and parts of the globe, and the chief points in the history of its diffusion there. It is true that something still remains unexplained; for instance, the immunity enjoyed by certain localities, the fact that an epidemic of the disease in Europe ceases entirely after the lapse of a certain time, and some other points. But it is no objection to well-established explanations, either here or in any other domain of human knowledge, that this or that point is still unexplained.

Other objections, current certainly till quite recent times, were aimed directly at the real character of Koch's Spirillum as the specific contagium of cholera. So far as they rested on the failure of attempts to communicate the infection with the pure Spirillum, they are set aside by the positive results now before us of similar attempts, supposing these to be trustworthy. On the other hand, they denied that Koch's organism was present exclusively in cases of Asiatic cholera. Finkler and Prior discovered a Spirillum extremely like it in the affection of the bowels known as indigenous cholera, cholera nostras. Lewis and, after him, Klein pointed to the comma-spirillum of the mucus of the mouth (see page 120 and Fig. 16, *d*), which is common in healthy human beings, and when seen in single specimens is also so like Koch's Spirillum that it might be considered to be identical. But further investigation has now removed all doubt concerning certain trustworthy distinctions, which come to light especially in cultures on a large scale, between these and other similar forms which cannot be enumerated here and Koch's organism; all attempts even to cultivate Lewis' Spirillum from the mouth as a saprophyte have as yet been attended with no positive result. Whether Finkler's and Prior's Spirillum may be the specific exciting cause of some other disease than Asiatic cholera cannot be considered here.

The most thorough-going objection, however, is that brought forward by Emmerich and supported by H. Buchner; these observers maintain that another than Koch's Bacterium is the specific contagium of cholera—a short non-motile rod-bacterium, which figures in literature under the provisional name of the Naples Bacillus.

Emmerich found his Bacterium in the gelatine-cultures which he employed in the examination of the bodies of persons who had died of cholera in Naples; fresh bits of the wall of the intestine, of the kidneys, and of other internal organs, together with blood from persons who had died of the disease or who were suffering from it, were placed with every precaution in the nutrient gelatine. The presence of the Bacterium in the above organs was concluded from the results of the culture; it grew in the culture, but it was not directly proved to have been present in the organs. The result of researches undertaken a year later in Palermo was, that no Bacterium could be shown to be present in the internal organs, liver, spleen, kidneys, or in the heart's blood in the majority of acute cholera-cases, nor was it found in the viscid exudation of the peritoneal cavity. In one case only was the Naples Bacillus obtained by culture from the liver of a patient. But it was now found in most cases in abundance in the contents of the stomach and intestine, though it must be added that there was some doubt as to the identity of the forms, and the decision on this point is reserved for further study. Moreover, the Naples Bacillus was obtained plentifully in most cases from the bronchial tubes and lungs. On the other hand, Emmerich and Buchner both bear witness to the practically constant presence of Koch's Spirillum in the intestine in the cases which they examined.

Emmerich also experimented on animals by inoculating them with pure material of his Bacillus, and obtained in this way positive results, the sickening with indubitable cholera-symptoms. But symptoms of this kind appear, as Virchow pointed out at the second Berlin Conference on cholera, when animals are inoculated with the most various kinds of substances which are putrid or contain Bacteria; they cannot therefore by themselves be regarded as decisive. It is true that this objection may also be made to the positive results mentioned above of infection with Koch's Spirillum. But all reported observations and results of experiment, the observations even of Koch's opponents, do at present agree in attesting the significance of his Bacillus as the specific contagium of cholera. Emmerich's results from Naples and Palermo, on the other hand, are not consistent with those of any other observer or with themselves; this detracts from the value of his experimental results in view of Virchow's observations. From the material before us the unprejudiced critic cannot, in my judgment, find any valid objection to the views of Koch and his school.

6. Among the diseases due to the action of Bacteria must be reckoned also traumatic infectious diseases, with their great variety of characteristic symptoms, affections also incident to child-bearing, and others connected with the formation of groups of ulcers, of abscesses of the skin and of the internal organs, of local skin-abscesses, boils, and ulcers, as well as more serious maladies (62). In these affections Bacteria-forms are found on the infected surfaces of the wounds, in pus, &c., and in all but a few cases, which for some special reasons are of an exceptional nature; and the eminent success of the antiseptic treatment of wounds introduced by Lister, the object of which is to prevent the approach of the organisms which set up decomposition, and to render them harmless, is an indirect proof according to our present views that these forms, as promoters of decomposition, are in causal connection with the affections in question.

This connection may be of two kinds. The contagium may cause local suppuration, formation of abscesses, &c., at the spot where it is found, either remaining in the wound which received it, or passing from it into the blood and with the blood into remote organs; or else unorganised poisonous bodies, ptomaines (see page 140), or similar substances are formed at the points of infection, as products of the vegetation of the contagium, and are then distributed in the blood and conveyed into the body, producing symptoms of poisoning in it. It is also conceivable that these two fundamentally distinct processes may occur in combination.

This subject can only be thus briefly noticed here; the details will be found in the extensive medical literature on the subject with which I am myself only imperfectly acquainted. I have chiefly followed Rosenbach's work, cited in note 62, on the contagia of traumatic infectious diseases. I will only add one remark, that a very noteworthy series of recent experiments are now offered for our study, in which inflammation, it is true, but no suppuration was caused by the application of the strongest chemical irritants of very different kinds, when the co-operation of Bacteria was excluded; Passet has however raised objections to the universal application of this principle.

As regards the Bacteria themselves of which we are speaking, several kinds have been observed. Rosenbach alone speaks of four different Bacilli or at least rod-forms, and especially of Micrococci, three kinds of which are common; the others need not be noticed here. The individual Micrococci cannot be certainly distinguished under the microscope; they are minute round bodies with a swarming movement only, and show no distinct formation of spores; but they are known from one another by their appearance when grouped together, and by the form and colour in which they show themselves in cultures on the large scale on the surface of agar-jelly. One genus which Billroth has named Streptococcus, has its cells united together in rows, in the manner of Micrococcus Ureae (see page 84). In the others the cells separate from the rows after division, and form aggregations which Ogston has compared with a bunch of grapes, and he has expressed the resemblance by the name Staphylococcus. One species of Staphylococcus forms orange-yellow, another white gelatinous expansions like the thallus of a Lichen on agar-jelly, and they are therefore known as Staphylococcus aureus and S. albus. If removed from abscesses and collections of pus and isolated in a pure culture, each of these Micrococci retains its characteristics unchanged; sometimes only one species occurs in these products of disease, sometimes the two are found together; from the accounts before us it would seem that Streptococcus and Staphylococcus aureus are the most common and the most destructive kinds. Rosenbach obtained positive results from inoculations and injections of pure culture-material obtained from men in several experiments on animals, that is, the introduction of the parasite caused fresh abscesses, though, if I rightly understand the accounts, large quantities of matter were employed in inoculation.

The above Bacilli and Micrococci are facultative parasites, and may be cultivated as saprophytes without difficulty and in abundance. No details are known with respect to their diffusion as saprophytes in nature, but experience of a more general kind makes it probable that these formidable foes exist everywhere, and especially in places of human resort; Passet has in fact found two of them (Staphylococcus aureus and S. albus) in dish-water and in putrid meat.

7. The Micrococcus which makes its way into the lymphatic ducts of the skin and is the contagium of erysipelas (63) is closely allied in respect of its shape and facultative parasitism to the forms of the preceding section, and is in effect a chain-forming Streptococcus. We owe our earlier knowledge of this organism to von Recklinghausen and Lukomski. Fehleisen has recently grown it in a pure state, and his inoculations with it were successful. The unpleasant though not dangerous affection of the skin of the hands, known as erythema migrans or finger-erysipelas, to which those are liable who have to handle raw meat, has been referred by Rosenbach to a Micrococcus (62).

8. A distinct species of Micrococcus, capable of cultivation and of being conveyed by inoculation with production of the characteristic disease is according to Sattler's investigation (64) the contagium of trachoma, the granulose inflammation of the conjunctiva of the human eye. It may be added that another disease of the conjunctiva of the eye, xerosis conjunctivae, is attributed to a small rod-bacterium as its exciting cause (65).

A Micrococcus forming short rows of cells in thick broad

gelatinous envelopes, and capable of cultivation in its characteristic form on gelatine, is said by Friedländer to have a strictly
specific effect as the contagium of acute fibrinous pneumonia
(66).

A specifically distinct Bacillus, nearly allied in every respect to
the Bacillus of tubercle, has been proved by Hansen's and
Neisser's investigations to be the exciting cause of leprosy in the
human species. The Bacillus discovered by Lustgarten, and
supposed to be the contagium of syphilis, is still a subject of
dispute (67). Other species of Bacilli, or at least of rod-forms,
approaching in their mode of life the Bacillus of anthrax, have
moreover been discovered, and many of them carefully studied
as the contagia of a series of diseases in the lower animals, such
as Koch's mouse-septicaemia, Koch and Gaffky's malignant
oedema (68), glanders (69), symptomatic anthrax (70), erysipelas in pigs (71), and Löffler's diphtheria in pigeons and in
calves (74).

9. The type of miasmatic infectious diseases is malaria,
intermittent fever with its kindred states (72). The infection is
confined to certain localities with a marshy soil and stagnant
water, and is not usually conveyed from one person to another.
It is natural to assume therefore, in accordance with other well-
known cases, that of anthrax for example, that an organism is
present in the soil and in the water of the malaria-district, and
that it causes the infection. Klebs and Tommasi Crudeli have
consequently examined specimens of soil and water from localities
where malaria abounds, and have found Bacteria in them in profusion, one especially, a rod-shaped species which forms filaments
and which they name Bacillus Malariae. They produced symptoms
of malarial fever, swelling of the spleen and intermittent fever in
different animals by injecting the Bacillus from these specimens
of the soil as well as from cultures. Cuboni and Marchiafava,
Lanzi, Perroncito, Ceci, and Ziehl have found Bacteria in blood
taken from the skin, veins, and spleen of persons suffering from
intermittent fever, especially in the cold stage of the attack.

Cuboni and Marchiafava have obtained in animals, into which
they had injected the blood of persons suffering from intermittent fever, symptoms which they considered to be those of
malaria-infection. I am not in a position to say how far these
symptoms observed in animals after infection may, or ought to
be regarded as sure signs of the presence of malarial fever.

On the other hand it is obvious that the injection of some
cubic centimetres of fluid containing particles of soil or of a
Bacillus-culture does not really correspond to an ordinary infection, in which a very minute portion of the contagium is in
every case absorbed, introduced by inoculation or inhaled.
And with respect to the descriptions given in the different publications of the Bacteria which were examined, we cannot be
sure whether one species of Bacterium was present in each case
or more than one, or whether the forms which one observer saw
in the blood were the same or of the same species as those
which others grew from soil-specimens. It therefore does not
seem to me that we have before us any precise determination of
the nature of the contagium or miasma vivum of malaria, but
that the question has now to be really attacked on the basis of
the former laborious investigations and with careful sifting of
their results. That these remarks, which appeared in the first
German edition of this book, were not without good foundation
is shown by the latest reports, especially those of Marchiafava
and Celli, in which it is stated that the contagium of malaria
is not a Bacterium, but a small amoeboid · organism which
penetrates into the red blood-corpuscles. We must hope for
clear and decisive investigations into this organism.

10. Our knowledge concerning the causal connection between
Bacteria or parasites generally, and typhoid fever and diphtheria
in men is also at present uncertain, notwithstanding Gaffky's
and Löffler's model investigations.

Typhoid fever is a distinctly miasmatic infectious disease,
which may sometimes become contagious. Causal relations
between its appearance and certain localities and the use of

impure water have long been clearly established. It is natural, therefore, as in the previous case, to suppose that a facultative parasite is the proximate cause of the disease. As early as 1871 von Recklinghausen found Bacteria, and especially colonies of Micrococcus, in the bodies of those who had died of this fever. Later investigations, given at length in Gaffky's publication (73), have resulted in reports of the occurrence of Bacteria and Fungi which do not always agree with one another. Gaffky has recently undertaken a thorough investigation of the subject, and has found in the internal organs, the mesenteric glands, spleen, liver, and kidneys of persons who have died of typhoid fever, as an almost constant phenomenon—in twenty-six out of twenty-eight cases,— a well-characterised endosporous Bacillus, and always the same kind. This species grows in characteristic form and abundantly on gelatine and blood-serum, and on potatoes exposed to the air. Gaffky, whose work should be consulted, states that in shape it is not unlike Bacillus Amylobacter (see page 100), but considerably smaller in size; each rod is about $2 \cdot 5 \mu$ in length, and the breadth is about one-third of the length. Contrary to the expectations which were justified by the unfailing and characteristic presence of the Bacteria in the bodies of the dead, Gaffky's extensive experiments in the infection of animals, and among them of monkeys, gave wholly negative results. The question of cause must therefore be considered to be at present undecided. How far more recent accounts of successful experiments of this kind will avail to alter this judgment I am not yet able to determine. It has also not yet been proved that the Bacilli of typhoid fever occur spontaneously outside the organism, especially in the water for drinking and for domestic use which is connected with epidemics of the fever.

11. We are indebted to Löffler (74) for extended and careful investigations into the nature of diphtheria. His work contains a detailed discussion of the statements of his predecessors, and should therefore be consulted. One well-known and characteristic symptom of diphtheria in the human subject is the forma-

tion of the white lining on the mucous membrane of the throat, especially on the tonsils, and it has been proved that by means of this lining the disease can be communicated to a healthy person. Enquiry was accordingly directed to this substance, and it was found to contain, along with a variety of accidental matters, first of all enormous accumulations of Micrococci, and secondly, small rods in many of the cases which were examined, but not in all of them, as was at first maintained by Klebs.

Löffler having ascertained the presence of these organisms subjected them to pure culture, and made experimental trial of their efficacy in the production of disease.

The Micrococcus in a pure culture forms chains which are very like those of erysipelas. It forces its way from the diphtheritic lining on the mucous membrane into the tissues of a person suffering from the disease, and passes through the lymph-vessels into the most dissimilar internal organs, where it forms nests. It behaved in a similar manner when introduced in the pure state into animals by inoculation, and produced some forms of disease, but did not give rise to the symptoms characteristic of diphtheria. We may therefore ascribe to this Micrococcus the power of producing morbid complications, but we cannot accept it as the specific contagium of diphtheria.

The rods thrive well on blood-serum, but are difficult of cultivation in other substrata. They are about as long as those of the Bacillus of tubercle and twice as thick, and are otherwise distinguished from this species by marks which cannot be fully described here. They are found collected together in heaps in the lining of the diphtheritic mucous membrane, in the layers beneath the surface. They have not been observed in the internal organs of sick persons. Introduced by inoculation into animals of the kinds usually employed for experiment, they produced symptoms very like those of diphtheria. Löffler therefore considers that the rods are the contagium of diphtheria in the human subject, though he is careful to draw attention to the objections to this conclusion.

12. In conclusion we must not omit to remark, that in a large number of infectious diseases, and these too of the most common occurrence, no one has as yet succeeded in discovering a distinct Bacterium or any other microscopic parasite as the cause of the particular disease, or the supposed discovery of such organisms is quite untrustworthy. This is true of diarrhoea, of typhus fever, of yellow fever, of whooping-cough, and of acute exanthemata of the skin, such as scarlet fever, measles, and small-pox in men and equivalent diseases in other animals. We practise vaccination, as is well known, as a protection against small-pox, and Pasteur applies his famous method for attenuating the contagium of hydrophobia, for protective inoculation and for curing the infected, but the organism which may possibly be the real contagium has hitherto at least escaped observation. It is not really necessary to repeat that Henle's postulates remain unchanged as against these negative results in the search for the contagium vivum.

XIV.

Diseases caused by Bacteria in the lower animals and in plants.

1. There is reason for assuming that Bacteria play a more important part as disease-producing parasites in cold-blooded as well as in warm-blooded animals, than is at present ascertained. What we do know at present is chiefly connected with insects (76).

The foul brood in bees, which may in a short time destroy the hives of a whole district, is the work of an (endosporous) Bacillus, B. melittophthorus, Cohn, in all probability the same species as B. alvei, which has been carefully studied by Cheshire and Cheyne.

The disease in silkworms known as flacherie is caused,

according to Pasteur, by a Bacillus and by a chain-forming Micrococcus, M. Bombycis, Cohn, which resembles M. Ureae (see page 84); these organisms are introduced with the food, and by decomposing it in the intestinal canal give rise first of all to derangement of the digestion, and then to the death of the insect, which first becomes inert, without appetite, and flabby, and soon succumbs to the disease. Its dead body is soft and soon turns a dark and dirty brown, putrefactive Bacteria make their appearance in it and it dissolves for the most part into discoloured stinking matter.

A number of other contagious and epidemic diseases in caterpillars of Lepidoptera have been recently referred by S. A. Forbes to the attacks of certain forms of Micrococcus and Bacillus.

There are two diseases among silkworms very distinct from the flacherie, which is the prevalent one at the present time, namely, muscardine or calcino, and the spotted disease or pébrine. Muscardine, known since the last century, was very destructive to the silkworm-culture in Europe during the first years of the present century, but is said to have almost entirely disappeared since 1855, while it continues to be of frequent occurrence among the caterpillars living in the wild state. It is caused, as has been fully shown, by a Fungus, and does not therefore belong to our present subject.

Pébrine (gattine, petechia, maladie des corpuscules) has been a known form of disease for some hundreds of years, and committed great ravages in Europe between 1850 and 1875. It received the name of spotted disease from the dark spots on the skin which make their appearance in the insect as it becomes dull and inert, and which are caused by the presence of a microscopic parasite, Panhistophyton ovatum, Lebert, Nosema Bombycis, Nägeli. The parasite is known under the form of small colourless, highly refringent bodies of irregularly ellipsoid shape, and not more than 0.4 μ in length, once termed the Cornalian bodies, which appear in our preparations either singly, or in pairs, or several connected together, and occur in all organs of the creature, and not in the caterpillar only, but also in the butterfly and even in the eggs, from which they may find their way again into the young caterpillar. They are also found in enormous quantities, filling the entire insect. Pasteur especially has shown that these bodies belong to a parasite, which penetrates into the insect, and multiplying at its expense produce the disease. If they are introduced with the food into the intestinal canal of a healthy caterpillar, they are subsequently found to have penetrated into the wall of the intestine, at first one by one, afterwards in greater numbers, and to have spread from thence into the other organs.

The same or similar bodies have been found by different observers in sundry other insects and articulated animals.

It would appear from this brief description that the Cornalian bodies resemble a small Bacterium, and specially a Micrococcus, and they have been regarded as such by many observers. Nägeli calls attention in his first communication to their relationship to Micrococcus aceti. This view rested on the similarity of form, and specially on the observation that they were frequently united together in pairs, since this fact was regarded as an indication of their multiplication by successive bipartition. Their bipartition was not directly observed at that time nor has it been observed subsequently, and it is obvious that union in pairs may be brought about in other ways. There was therefore experimental proof that the bodies did multiply, but how they multiplied was not known. Then Cornalia, Leydig, Balbiani, and also Pasteur put forward another view in opposition to the theory that the bodies were Micrococci; they supposed them to be totally distinct from Micrococcus and Bacterium, and to be Psorospermiae, that is, states of peculiar low organisms, Sporozoa or Sarcosporidia. Metschnikoff has recently and distinctly confirmed this view; he states briefly that the parasite of pébrine consists of protoplasmic bodies having the power of amoeboid movement, after the manner, that is, of the colourless blood-cells described on page 135, and subsequently becoming lobed, and that the Cornalian bodies are produced in them by endogenous formation. It follows from analogy with other better-known Sporozoa, that the Cornalian bodies would be spores, and that their germination gives rise to the amoeboid protoplasmic bodies in which fresh spores are formed in large numbers. The extreme delicacy of such amoeboid protoplasmic bodies as these sufficiently explains why it was so long before they were clearly made out, especially when they had penetrated into and were enclosed in the tissue of the insect's body which is also protoplasmic.

Hence the parasite of pébrine must also be excluded from consideration in a treatise on Bacteria; nor would so much have been said about it here, if it did not serve to show in a very instructive manner, not only that very minute parasites which are not Bacteria are the contagium in infectious diseases in animals, but that we may be dealing with organisms of a quite different kind, conformation, and mode of life from Bacteria, even when the figures before us are very like Bacteria and are easily mistaken for them.

2. Lastly, parasitic Bacteria do not often appear, according to our present experience, as the contagia of diseases in plants (77). Most contagia of the many infectious diseases in plants belong to other groups of animals and plants, the larger number to the Fungi proper, as was observed on pages 146, 147.

Among cases of the kind we may mention first the yellow disease studied by Wakker, which destroys hyacinth-plants. Wakker found that the most characteristic symptom in this disease is the appearance of a rod-shaped Bacterium, 2.5 μ long and a fourth or half as broad, which aggregates into slimy yellow masses filling the vessels and tissue of the vascular bundles in the bulb-scales during the time when vegetation is dormant. At flowering time these masses ascend also into the leaves, where they are not confined to the vascular bundles, but make their way from them into the intercellular passages and into the cells of the parenchyma, stopping up the passages and destroying

N

the cells, and ultimately emerging through the bursting epidermis. Attempts to communicate the disease by inoculation have not yet been successful, nor has the life-history of the Bacterium been at present thoroughly worked out.

J. Burrill, of Urbana, in the state of Illinois, describes a disease in pear-trees and apple-trees, known by the indefinite name of 'blight,' and attributes it to the attack of a Bacterium, a rather elongated Micrococcus, M. amylovorus, Burr., the cells of which are about 1 μ in length. The disease, which destroys the bark, is at first narrowly localised, but may spread and form a ring round the branch or stem which it attacks, and may then prove fatal to it. Burrill found that the Micrococcus had penetrated into the cells of the diseased part, and that as it developed the normal contents of the cells, especially the starch, disappeared, while ' carbonic acid, hydrogen and butyric acid' were formed. Numerous attempts at inoculation by introducing the Micrococcus into small incisions or punctures in the bark of healthy pear-trees and apple-trees resulted in the communication of the disease. Arthur has confirmed Burrill's observations and given further proof that his Micrococcus is a facultative parasite, producing effects peculiar to the species. This disease of pear-trees, as far as I am aware, is either unknown in Europe, or has not been investigated.

From some very brief statements of Burrill it would appear that diseases caused by Bacteria also occur in the peach-tree, the Italian poplar, and the American aspen.

Prillieux gives a short description of a change which sometimes takes place in grains of wheat; this change which is recognised by the rose-red colour which it produces, advances *pari passu* with the development of a Micrococcus, which destroys the starch-grains, the glutinous contents of the peripheral cell-layers, and to some extent the cell-membranes as well. Disorganising operations of the Micrococcus are thus distinctly disclosed. Its real significance as an exciting cause of disease cannot be certainly determined from the few published statements, it may perhaps only play a secondary part as a saprophyte in consequence of injuries produced by other causes.

The latter supposition is partly founded on the phenomenon of wet rot in potatoes, which has been closely examined by Reinke and Berthold. It appears from the observations of these authors, that the proximate cause of the phenomenon is the development of Bacteria; Bacillus Amylobacter is shown by the descriptions to be present, and perhaps some other forms. The wet rot usually attacks tubers which have been previously sickly, that is, have been partly destroyed by a purely parasitic Fungus, Phytophthora infestans. The rot does indeed attack the tissue which had been spared by the Fungus and is still alive, but it nevertheless is only a secondary phenomenon. At the same time the rot appears in potatoes which have not suffered from Phytophthora, though this is exceptional; and the above-mentioned observers succeeded in producing wet rot in healthy potato-tubers by inoculating them with their Bacteria. This agrees with a recent experiment of van Tieghem, who succeeded in entirely destroying living tubers by the introduction of Bacillus Amylobacter into their inner substance, and by keeping them at the same time at the high temperature of 30° C. The same results were obtained with the seeds of beans, stems of Cacti, &c. In other words, these facts show that saprophytic Bacteria may also, under special conditions, attack the tissues of living plants as facultative parasites, produce disease in them and destroy them. But this only confirms what was said above, that Bacteria are not objects of great importance as contagia of diseases affecting plants.

CONSPECTUS OF THE LITERATURE
AND NOTES.

1. FOR the general literature of the Bacteria see de Bary, Comparative Morphology and Biology of the Fungi, Mycetozoa, and Bacteria, Clarendon Press, 1887, and W. Zopf, Die Spaltpilze, 3rd edition, Breslau, 1884.— The works of Pasteur, F. Cohn, v. Nägeli, van Tieghem, R. Koch, Brefeld, A. Prazmowski, Fitz, must be specially mentioned here as laying generally the foundations of our knowledge; they are cited in the works above-named, and some of them again below.—Duclaux, Chimie biologique, Paris, 1883, gives an elegant account of the views and methods of the French school, and especially of the school of Pasteur; F. Hueppe, Die Methoden d. Bacterienforschung (3rd ed., 1886), gives hints for the conduct of investigation according to the method perfected especially by R. Koch.—Writers on General Morphology and Classification are F. Hueppe, Die Formen d. Bacterien &c., Wiesbaden, 1886; J. Schröter in the Kryptogamenflora v. Schlesien, ed. F. Cohn, Bd. III, 2 Lieferung, pp. 136-172. This work gives a good classification and description of most known forms; it reached me while the present work was in the press, and it was hardly possible for me to make any use of it.—Of the many general Text-books of Bacteria of recent date may be mentioned: C. Flügge, Fermente u. Mikroparasiten, in v. Pettenkofer and v. Ziemssen, Handb. d. Hygieine (the second edition with the title, Die Mikroorganismen, Leipsic, 1886, came out while this work was being printed); E. M. Crookshank, Introduction to practical Bacteriology, London, 1886; and a copious Text-book of the Bacteria of Disease, by Cornil et Babes, Les Bactéries et leur rôle dans l'anatomie et l'histologie pathologiques des maladies infectieuses, 2nd ed., Paris, 1886. With these may be coupled the Jahresbericht ü. d. Fortschritte d. Lehre v. d. pathogenen Mikroorganismen of P. Baumgarten, Erster Jahrg. 1885, Braunschweig, 1886, of which I have made frequent use, and to which I here refer the reader once for all for the more recent special literature.

2. Nencki u. Schaffer, Journ. f. pract. Chemie, Neue Folge, XX.— Nencki, Berichte d. D. Chem. Ges. Jahrg. XVII, p. 2605.

3. Leeuwenhoek, Experimenta et contemplationes, Delft, 1695, especially p. 42, on Bacteria-forms from saliva.

4. F. Cohn, Unters. ü. Bacterien, in Beitr. z. Biologie d. Pfl., continued since 1872 (I, Heft 2, p. 127).

5. C. G. Ehrenberg, Die Infusionsthiere als vollk. Organismen, Berlin, 1838.

6. Billroth, Coccobacteria septica, Berlin, 1874.

7. v. Nägeli, Die niederen Pilze, München, 1877.

8. Hornschuch in Flora, Regensburg, 1848.

9. v. Nägeli, Niedere Pilze, 1877, p. 21.

10. F. Hueppe, Unters. ü. d. Zersetzgn. d. Milch., in Mittheil. aus d. K. Gesundheitsamt, II, 1884.

11. C. Vittadini, Della natura del calcino, in Giorn. Istitut. Lombardo, III (1852).

12. E. Klebs, Beitr. z. Kenntn. d. Mikrokokken, in Arch. f. exp. Pathologie, I (1873).

13. Pasteur, Examen de la doctrine des générations spontanées in Ann. de Chimie, sér. 3, LXIV. See also Ann. d. sc. nat., Zoologie, sér. 4, XVI. —Rosenbach has given an account of Meissner's excellent researches in Deutsche Zeitschr. f. Chirurgie, XIII, p. 344. For more recent works and discussions, see in Baumgarten's Jahresber.

14. R. Koch in Mittheil. aus d. K. Gesundheitsamt, I, p. 32. See also Hesse, in the same publication, II, 182.

15. Annuaire de l'observatoire de Montsouris, since 1877, and especially since 1879.

16. Virgil, Georgics, IV, 281.

17. A. Béchamp, Les Microzymas dans leurs rapports avec l'hétérogénie, l'histiogénie, la physiologie et la pathologie, Paris, 1882. In this volume Béchamp has brought together the views successively entertained by him and published in the Comptes rendus of the Paris Academy.

18. A. Wigand, Entstehung u. Fermentwirkung d. Bacterien, Marburg, 1884.

19. O. Brefeld, Botan. Untersuch. ü. Schimmelpilze, IV.

20. E. Eidam in Cohn's Beitr. z. Biol. d. Pflanzen, I, Heft 3, p. 208.

21. A. Fitz in Ber. d. Deutsch. Chem. Ges., nine papers in the years 1876–84.

22. Frisch in Sitzungsber. d. Wien. Acad., Mai, 1877.

23. P. van Tieghem in Bull. de la Soc. Bot. de France, XXVIII (1881), p. 35.

24. E. Duclaux, Études sur le lait, in Ann. de l'Instit. Nat. Agronomique, No. 5, Paris (1882), pp. 22–138.

25. E. Duclaux, Chimie biolog., in Encyclop. Chimique, publiée par M. Frémy, IX, Paris, 1883.

26. W. Engelmann in Bot. Ztg., 1882, p. 321.

27. v. Nägeli, Ernährung d. niederen Pilze, in Sitzungsber. d. Münchener Acad., Juli, 1879.

28. v. Nägeli, Unters. ü. niedere Pilze, in Unters. aus d. Pflanzenphysiol. Inst. z. München, München, 1882.

29. W. Engelmann, Bacterium photometricum, in Unters. aus d. Physiol. Laboratorium z. Utrecht, 1882.

30. Cohn u. Mendelssohn in Beitr. z. Biol. d. Pflanzen, III.

31. W. Engelmann in Bot. Ztg., 1881, p. 441.

32. W. Pfeffer, in Unters. a. d. Bot. Inst. z. Tübingen, I, Heft 3.

33. J. Tyndall in Phil. Trans. of the Royal Soc., London, 166 (1876), 167 (1877). The latter paper especially contains the statements with respect to fractionating sterilisation.

34. J. Wortmann in Zeitschr. f. physiol. Chemie, VI, p. 287.

35. I still keep to the classification and nomenclature founded on Cohn's divisions. It is better to treat what is imperfect as imperfect than to give it the appearance of being perfect, and so cheat oneself and the beginner into belief in a state of things which does not really exist. This might also be said in criticism of some of the most recent attempts at improvement, but further discussion of these would be out of place here. We have not yet accomplished that which is an indispensable condition for any real advance in classification, namely, a tolerably continuous morphological examination of the individual species,—I do not mean of all existing species, but only of such as at present pass for being ‘described,’—whether collective or ‘bad’ species. Such an examination might be made with the material which we at present possess, but this has not yet been done.

36. W. Zopf, Zur Morphol. d. Spaltpflanzen, Leipsic, 1882 ;—Id., Entwicklungsgeschichtl. Unters. ü. Crenotbrix polyspora, die Ursache d. Berliner wassercalamität, Berlin, 1879 ; — Id. in Monatsber. d. Berliner Acad., März 10, 1881.

37. E. Warming, Om nogle ved Danmarks Kyster levende Bacterier, in Vidensk. Meddelelser fra d. naturhist. Forening, Kjöbenhavn, 1875.—A. Engler, Die Pilzvegetation d. weissen od. todten Grundes d. Kieler Bucht, in Ber. d. Commiss. z. Erforschung d. deutschen Meere, IV.

38. P. van Tieghem, Sur la fermentation ammoniacale, in Compt. rend. LVIII (1864), p. 211.—v. Jacksch in Zeitschr. f. physiolog. Chemie, V (1881), p. 395.—Leube, Ueber d. ammoniakal. Harngährung. in Virchow's Archiv, C, p. 540.

39. Schlössing u. Müntz in Compt. rend. LXXXIV, p. 301 ; LXXXIX, pp. 91, 1074.

40. Pasteur in Compt. rend. LIV, p. 265 ; LV, p. 28 ;—Id., Études sur le vinaigre, Paris, 1868.

41. v. Nägeli, Theorie d. Gährnng, München, 1879.

42. E. C. Hansen, Beitr. z. Kenntn. d. Organismen welche, in Bier u. Bierwürze leben, in Meddelelser fra Carlsberg Laboratoriet, I, Kopenhagen, 1882.

43. Pasteur in Compt. rend. LII, p. 344.

44. P. van Tieghem, Leuconostoc, in Ann. d. sc. nat. sér. 6, VII.

45. F. Hueppe, Unters. ü. d. Zersetzung d. Milch durch Mikroorganismen, in Mittheil. a. d. k. Reichsgesundheitsamt, II, 309, with a full account of the history and literature.

46. E. Kern, Ueber ein Milchferment aus dem Kaukasus, in Bot. Ztg.,

1882, p. 264, and in Bullet. de la Soc. Imper. d. Nat. de Moscou, 1881.—See also F. Hueppe, Ueher Zersetzungen d. Milch, &c., in Börner's Deutscher med. Wochenschrift, 1884, No. 48.—W. Podwyssotzki (Sohn), Kefyr, kankasisches Gährungsferment u. Getränk, &c., translated from the Russian of the 3rd Edition by Moritz Schulz, St. Petersbg., 1884.—W. N. Dimitrijew, Kefir oder Kapir, echtes Kumyiss aus Kuhmilch, translated from the German by E. Rothmann, St. Petersbg., 1884.—Alexander Levy, Die wahre Natur des Kefir, in Deutsche medicinal Zeitung, 1886, p. 783. Levy's manner of proceeding is to add one part of ordinary sour milk to 8-10 parts of cold boiled milk, and then to shake the mixture at a temperature of about 12°C.

47. P. van Tieghem, Sur le Bacillus Amylobacter, &c., in Bull. Soc. Bot. de France, XXIV (1877), p. 128;—Id., Sur la fermentation de la cellulose, in Bull. Soc. Bot. de France, XXVI (1879), p. 25.

48. A. Prazmowski, Untersuch. ü. Entwicklungsgesch. u. Fermentwirkung einiger Bacterien-Arten, Leipzig, 1880.

49. Vandevelde, Studien z. Chemie d. Bacillus subtilis, in Zeitschr. f. physiolog. Chemie, VIII, 367.

50. Bienstock, Ueber d. Bacterien d. Faeces, in Zeitschr. f. klin. Medicin, VIII.

51. Cohn, Beitr. I, Heft 2, p. 169.—G. Hauser, Ueber Faulnissbacterien u. deren Beziehung z. Septicaemie, Leipzig, 1885. I have made little use of this work in the text, because I am unable sufficiently to understand its morphology without personal investigation. The whole of the phenomena of 'pleomorphism,' on the strength of which the writer designates his forms by the special generic name Proteus, an unacceptable one in any case, appear to me, on comparing the figures, scarcely to deserve the appellation. But the figures do not help us in determining the morphological relations, at least if we go beyond the habit of the groups; I have taken fruitless pains to arrive by their means at a clear idea of the conformation of the Spirilla-forms which he describes. Photography is no doubt a useful assistant in microscopical studies, and I have used it myself for twenty-five years; but there are limits to its capabilities, and the details required in this case are not given in Hauser's figures. With all due acknowledgment of the advance made in Hauser's work, I say what I have now said in order to justify my casual treatment of it.

52. H. Nothnagel, Die normal in d. menschl. Darmentleerungen vorkommenden niedersten pflanzl. Organismen, in Zeitschr. f. klin. Medicin, III (1881).—Kurth, Bacterium Zopfii, in Bot. Ztg. 1883, 369.—Miller, Ueber Gährungsvorgänge im Verdauungstractus, &c., in Deutsche med. Wochenschrift, 1885, No. 49.—W. de Bary, Beitr. z. Kenntn. d. niederen Organismen im Mageninhalt, in Archiv f. experim. Pathologie u. Pharmacologie, XX.

53. The literature of Sarcina has been carefully collected by Falkenheim in his paper Ueber Sarcina, published in Archiv f. experim. Pathologie, XIX, p. 339.

The species of Sarcina which I am able to distinguish are characterised as follows :

a. Sarcina ventriculi, Goodsir (see Fig. 14 of this book).   Large cube-shaped packets, with very many members, that is, consisting usually of 64–4096 cells, the single packet brownish-gray in transmitted light under the microscope, of a dirty greyish-white colour when seen in mass by the naked eye in reflected light.   Single cells, 3–4 μ in size, stained a dirty violet by Schulze's solution.

In large quantities of material from the human stomach I generally find two distinct forms side by side, a large-celled one which is represented in Fig. 14, and another in which the cells are smaller (2 μ) and less clearly translucent.   I can give no information respecting the genetic relations of the two forms.

b. Sarcina Welckeri, Rossmann (Welcker in Henle's u. Pfeffer's Zeitschr. f. rat. Med. V, 199).   Small cube-shaped packets, containing at most 64 cells, quite colourless.   Single cells about 1 μ in size, membranes not coloured by Schulze's solution, protoplasm coloured yellow by the same solution.   It occurs in the urinary bladder of the living human subject ; repeatedly found in patients.   I know the species as coming from a young man, who, according to his own observation, voided them with the urine for more than twelve months, in varying, often in very large, quanti-ties.   The urine is usually abnormally rich in phosphates ; the patient is in other respects in good health.   The Sarcina did not grow with me on gelatine and agar ; once doubtfully and not in any abundance in sterilised urine at a temperature of 35°C.   Most of the attempts to cultivate it in urine gave negative results.

c. Sarcina flava.   Small packets, containing from 16 to 32 cells, con-nected together in the majority of cases either into large regular cubes or into irregular heaps ; the single packet colourless in transmitted light under the microscope, of a beautifully bright yellow colour when seen in mass in re-flected light.   Single cells 1–2 μ in size.   Reaction with iodine as in Sarcina Welckeri.   Grows well on gelatine, which it quickly liquefies, on agar and on other substances.   Appears to be comparatively abundant ; is cultivated in laboratories, and described without statement as to the source from which it is obtained.   See Crookshank, Introd. to practical Bacteriology, London, 1886.   My material comes from the Pathological Institute at Greifswald, where the yellow Sarcina made its appearance spontaneously as an isolated colony in a gelatine culture of vomited matter, in which Sarcina ventriculi had not been found.   It is at all events distinguished from S. lutea of Schröter by its power of liquefying gelatine.

d. Sarcina minuta, provisionally a new species (see Fig. 15 of this book).   Small cube-shaped packets, formed of 8–16 cells, connected together in the majority of cases in irregular heaps, less frequently in larger cubes ; quite colourless.   Single cells about 1 μ in size.   Reactions with iodine as in Sarcina Welckeri.   Made its appearance once spontaneously in a culture of sour milk on a microscopic slide. Grows well but slowly on gelatine and in a saccharine solution with extract of meat, forming in the solution the regular cube-shaped packets, on the gelatine the irregular heaps. Closely resembling Sarcina Welckeri under the microscope.

ε. Sarcina fuscescens. Small cube-shaped packets of 8–64 cells, readily separating into smaller groups (tetrads) or into single cells. Single cells about 1·5 *μ* in size. Reactions with iodine as in Sarcina Welckeri. Forms small brownish scales or mould-like films on the substrata mentioned below. This form was obtained by Falkenheim in gelatine cultures of the contents of a human stomach containing Sarcina ventriculi, with which, however, there was no apparent genetic connection. It grew in connected packet-form on infusion of hay; culture on other ordinary nutrient substrata (gelatine, potatoes, &c.) was accompanied by the separation mentioned above into single cells and smaller cell-groups. The above description is taken from Falkenheim's paper on Sarcina in Archiv f. experim. Pathologie, XIX, p. 339; the form or species is not known to me by personal observation.

To the above must be added the distinct and described species, Sarcina intestinalis, Zopf (Spaltpilze, p. 55 of the third edition), from the intestinal canal of the domestic fowl; S. lutea, Schröter (Krypt. Fl. v. Schlesien), a saprophyte which makes its appearance in Fungus-cultures; and Schröter's S. rosea and S. paludosa, which live in bog-water.

Other forms, belonging apparently to the genus Sarcina, are mentioned by J. Eisenberg (Patholog. Diagnostik), but they are without any special description, and cannot therefore be compared with the preceding species. A ‘Sarcina in the mouth and lungs’ has been considered by H. Fischer, in the Deutsches Archiv f. klin. Medicin, XXXVI, p. 344; but it is not even clear from his description whether the cells or divisions of cells are arranged according to two or three dimensions; we are therefore unable to compare this supposed form with the rest of the species.

The names Sarcina littoralis, Oersted, S. hyalina, Kützing, and S. Reitenbachii, Caspary (also misprinted Reichenbachii), have been copied into the literature of the Bacteria. The proximate source of these names is Winter's Pilzflora v. Deutschland, &c. Merismopoedia littoralis, Rabenhorst, M. hyalina, Kützing, M. Reitenbachii, Casp., have thus been placed in the genus Sarcina; this appears to me to be incorrect, because, as the name Merismopoedia implies, the cells in accordance with the directions of their divisions, form not many-layered packets, but tables of a single layer, and because it is uncertain, at least in the case of the two last forms, whether, like other species of Merismopoedia, M. punctata for example, they do not contain chlorophyll or phycochrome. These forms, therefore, do not belong to this place; like other species of Merismopoedia they live in bogs and in sea-water.

54. Rasmussen, Ueber d. Cultur v. Mikroorganismen aus d. Speichel (Spyt) gesunder Menschen; Dissert. Kopenhagen, 1883. Known to me only from an abstract in the Bot. Centralblatt, 1884, XVII, p. 398.

55. W. Miller, Der Einfluss d. Mikroorganismen auf d. Caries d. menschl. Zähne, in Archiv f. exp. Pathol. XVI, 1882 ;—Id., Gährungsvorgänge im menschlichen Munde in Beziehung zur Caries d. Zähne, &c., in Deutsch. med. Wochenschrift, 1884, No. 36.—T. Lewis, Memorandum on the comma-shaped Bacillus, &c., in the Lancet, Sept. 2, 1884.

56. For the earlier literature of Anthrax (up to 1874) see O. Bollinger, in Ziemssen's Handb. d. speciellen Pathologie u. Therapie, 3, and the rich material in Oemler, Experimentelle Beitr. z. Milzbrandfrage in Archiv f. Thierheilk, II–VI. For the first discovery of the Bacillus, see Rayer in Mémoires de la Soc. de Biol. II, 1850, p. 141 (Paris, 1851).—Pollender in Casper's Vierteljahrsschr. VIII, 1855. Of the very numerous works of a more recent date may be mentioned : Pasteur in Compt. rend. LXXXIV (1877), p. 900 ; LXXXV (1877), p. 99 ; LXXXVII (1878), p. 47 ; XCII (1881), pp. 209, 266, 429.—R. Koch, Die Aetiologie d. Milzbrandes, in Cohn, Beitr. z. Biol. d. Pfl. II, 277 ;—Id. in Mittheil. a. d. Reichsgesund-heitsamt I, and II in conjunction with Gaffky and Löffler.—H. Buchner in Unters. aus d. Pflanzenphysiol. Inst. z. München, 1882.—Chauveau in Compt. rend. XCI (1880), p. 680 ; XCVI (1883), pp. 553, 612, 678, 1471 ; XCVII (1883), pp. 1242, 1397 ; XCVIII (1884), pp. 73, 126, 1232.—Gibier in Compt. rend. XCIV (1882), p. 1605.—E. Metschnikoff, Die Beziehung d. Phagocyten z. d. Milzbrand-Bacillen, in Virchow's Archiv, XCVII, 1884.—A. Prazmowski in Biolog. Centralblatt, 1884.

57. Pasteur in Compt. rend. XC (1880), pp. 239, 952, 1030 ; XCII (1881), p. 426.—Semmer, Ueber d. Hühnerpest, in Deutsche Zeitschr. f. Thier-medicin, IV (1878), p. 244. The disease described by Perroncito, in Archiv f. wiss. u. pract. Thierheilk. V (1879), p. 22, must be of another kind.—Kitt, Exper. Beitr. z. Kenntn. d. epizootischen Geflügeltyphoids, in Jahresber. d. k. Thierarzneischule in München, 1882–1884, p. 62, Leipzig, 1885. This excellent work contains an exact account firstly of the form of the Micro-coccus and its behaviour in cultures, and secondly of experimental investiga-tions into infection and the phenomena of the disease which follows upon the infection. Though it confirms Pasteur's statements in some points, it tends at the same time to throw doubt on others, by making it highly probable that Pasteur did not work with pure material, free from other Bacteria. It makes no mention of the state of stupor. When the animals, especially fowls, succumbed to the disease, death usually occurred in twenty-four hours and suddenly. It is possible that Pasteur examined one infectious disease and Kitt another; after the latter's thoroughly trustworthy repre-sentations, Pasteur's statements certainly require critical examination. If on the whole I have left them for the present in the text, I have done it with this reservation, and especially on account of their importance for the development of the doctrine of contagia viva, which importance they retain even if they prove not to be entirely correct.

58. The following works may be consulted in special connection with this section, but the reader is at the same time referred to the medical

literature generally, and especially to Liebermeister's Introduction to Infectious Diseases in Ziemssen's Handbuch, II. J. Henle, Patholog. Unters. I, Berlin, 1840.—The works cited in note 56.—de Bary, Die Brandpilze, Berlin, 1853 ;—Id., Recherches sur le développement de quelques champignous parasites iu Ann. d. sc. nat. (Botanique), sér. 4, XX.—Frank, Die Krankh. d. Pflanzen, Breslau, 1880.—de Bary, Morphol. and Biol. of the Fungi, &c., Clarendon Press, 1887 ;—Id., in Jahresber. ü. d. Leistungen u. Fortschr. d. Medicin, herausg. von Virchow u. Hirsch, II (1867), Abth. 1, p. 240 ;—Id., Ueber einige Sclerotienkrankheiten, &c., in Bot. Ztg., 1886.— v. Recklinghausen, in Berichte d. Würzburger phys.-med. Ges., 1871.— E. Klebs in numerous papers, especially in Archiv f. experiment. Pathol. n. Pharmacol. I (1873).—v. Nägeli, Die niederen Pilze, München, 1877.

59. O. Obermeier, in Berliner klin. Wochenschr. 1873.—Cohn, Beitr. z. Biol. d. Pfl. I, Heft 3, p. 196.—v. Heydenreich, Unters. ü. d. Paras. d. Rückfalltyphus, Berlin, 1877.—R. Koch in Mittheil. d. Reichsgesundheitsamts, I.

60. R. Koch, Die Aetiologie d. Tuberculose, in Mittheil. d. Reichsgesundheitsamts, II.—Malassez et Vignal, Tuberculose zoologique in Compt. rend. XCVII (1883), p. 1006 ; XCIX (1884), p. 203.

61. Neisser in Centralblatt f. d. med. Wissensch., 1879, and in Deutsche Med. Wochenschr., 1882, No. 20.—Bockhardt, Beitr. z. Aetiologie u. Pathologie d. Harnröhrentrippers, in Sitzungsber. d. phys.-med. Ges. z. Würzburg, 1883, p. 13.—E. Bumm, Der Mikroorganismus d. Gonorrhoischen Schleimhaut-Erkrankungen, Wiesbaden, 1885.—See also Nägel in Jahresbericht, &c. d. Ophthalmologie.

62. From the very copious literature of wound-infection, I cite here only F. J. Rosenbach, Mikroorganismen bei d. Wundinfectionskrankheiten d. Menschen, Wiesbaden, 1884.—J. Passet, Unters. ü. d. eitrige Phlegmone d. Menschen, Berlin, 1885. In these books and in Baumgarten's Jahresbericht will be found further information concerning the literature of the subject.

63. v. Recklinghausen u. Lukomski in Virchow's Archiv, LX.—Fehleisen in Deutsche Zeitschr. f. Chirurgie, XVI, p. 391.—Koch in Mittheil. d. Reichsgesundheitsamts, I.

64. Sattler, Die Natur d. Trachoms, &c., in Ber. ü. d. Versammlung d. ophthalmol. Ges. z. Heidelberg, 1881, p. 18; 1882, p. 45.—Michel, in Sitzgsber. d. Würzb. phys.-med. Ges., 1886.

65. Kuschbert u. Neisser in Deutsche med. Wochenschr., 1884, No. 21. —Schleich, Zur Xerosis conjunctivae, in Nägel's Mitth. aus d. ophth. Klinik z. Tübingen, II, p. 145.

66. C. Friedländer, Ueber d. Schizomyceten b. d. acuten fibrinösen Pneumonie, in Virchow's Archiv, LXXXVII (1882), p. 319 ;—Id. in Fortschritte d. Medic. I, 1883. For more recent confirmatory statements, see in Baumgarten's Jahresbericht.

67. On leprosy, see Neisser in Ziemssen's Handb. d. spec. Pathol. u.

Therapie, XIV. For more recent works on leprosy, by Unna and others, see in Baumgarten's Jahresbericht, and also for discussions on the Bacillus of Syphilis.

68. Mittheil. d. Reichsgesundheitsamts, I.

69. Bollinger in Ziemssen's Handb. III.—Löffler u. Schütz in Deutsche med. Wochenschrift, 1882, p. 707.—O. Israel in Berliner klin. Wochenschr., 1883, p. 155.—Kitt in Jahresber. d. k. Thierarzneischule München, Leipzig, 1885.

70. Bollinger n. Feser in Deutsche Zeitschr. f. Thiermedicin, 1878–1879. —T. Ehlers, Unters. ü. d. Rauschbrandpilz ; Dissert. Rostock, 1884.—Kitt in Jahresber. d. k. Thierarzneischule München, Leipzig, 1885.

71. E. Klein in Virchow's Archiv, XCV (1884), p. 468.—For Löffler, Lydtin, Schottelius, and Schütz, see Baumgarten's Jahresbericht, 1884, p. 101.

72. Klebs und Tommasi Crudeli, Studien ü. d. Ursache d. Wechselfiebers n. ü. d. Natur d. Malaria, in Archiv f. experim. Pathol. XI.—Cuboni u. Marchiafava in Archiv f. experim. Pathol. XIII. — Ceci in Archiv f. experim. Pathologie, XV and XVI.—Ziehl in Deutsche med. Wochenschrift, 1882, p. 647.—Marchiafava u. Celli, N. Unters. ü. d. Malaria-Infection, &c., 1885.—See Baumgarten's Jahresbericht, 1885, p. 153. I know these authors only from this report. Laverans' book, cited in it, I do not know. A criticism of the statements of the Italian observers, as they lie before me, would be out of place here ; I would only counsel them to study attentively the zoologico-botanical literature of that portion of natural history with which their writings are occupied, before they deal with it so independently, and employ terms like 'plasmodium,' for the word is applied with the utmost precision to a portion of developmental history which the authors have never observed in the cases which they are discussing.

73. Gaffky, Zur Aetiologie d. Abdominaltyphus, in Mitth. aus d. k. Reichsgesundheitsamt, II, 372, where the literature is given at length.

74. Fr. Löffler, Unters. ü. d. Bedeutung d. Mikroorganismen für d. Entstehung d. Diphtherie beim Menschen, bei d. Taube u. beim Kalbe in Mitth. aus d. k. Reichsgesundheitsamt, II, p. 421.

75. J. M. Klob, Pathol. anatom. Studien ü. d. Wesen d. Choleraprocesses, Leipzig, 1867.—R. Koch in Berliner klin. Wochenschrift, 1884, Nos. 31–32 a;—Id. in Verhandl. d. zweiten Conferenz z. Erorterung d. Cholerafragen in Berlin. klin. Wochenschrift, 1885, No. 37 a.—E. van Ermengem, Recherches sur le Microbe du Choléra Asiatique, Paris et Bruxelles, 1885.—F. Hueppe, Ueber d. Dauerformen d. sogen. Kommabacillen, in Fortschritt d. Medicin, III, 1885, No. 19, and in Deutsche med. Wochenschrift, 1885, No. 44.— For Nicati u. Rietsch, Doyen, Watson Cheyne, Babes, and others, see the citations in Baumgarten's Jahresbericht, 1885.—Finkler u. Prior in Tagebl. d. sieben u. funfzigsten Ver. Deutsch. Naturf. u. Aerzte z. Magdeburg, p. 216;—Id., Forschungen ü. Cholerabacterien im Ergänzungsheft z. Central-blatt f. allg. Gesundheitspflege, I.—Emmerich, Vortr. im Aerztl. Ver. z.

München reported in Berl. klin. Wochenschrift, 1885, No. 2 ;—Id. in Arch. für Hygieine, III.—H. Buchner in Arch. für Hygieine, III.—Emmerich u. Buchner, Die Cholera in Palermo, in Münchener med. Wochenschrift, 1885, No. 44.—v. Sehlen, Bemerkungen ü. d. Verhalten d. Neapler Bacillen in d. Organen, &c., in Münchener med. Wochenschrift, 1885, No. 52.— J. Ferran, Die Morphologie d. Cholera-Bacillus u. d. Schutz-Cholera-Impfung, by Dr. Max Breiting, after Dr. Ferran, in Deutsch. medicin. Zeitung, IV (1885), p. 160, and Ueber d. Morphol. d. Komma-Bacillus in Zeischr. f. klin. Medicin, edited by Leyden, Bamberger, and Nothnagel, IX (1885), p. 375, t. 11.—Cholera, Inquiry by Doctors Klein and Gibbes, and Transactions of a Committee convened by the Secretary of State for India in Council, 1885.—See also E. Klein in Proceedings of the Royal Soc. of London, XXXVIII, No. 236, p. 154.—T. Lewis in the Lancet, Sept. 2, 1884.

The account given in the text is founded on the literature quoted above, and has been modelled chiefly on van Ermengem's excellent book. The botanical description of the Spirillum of cholera is also founded partly on the numerous descriptions and figures which we possess, partly on my own examination of Finkler's and Prior's form. I was limited to this, because my request for a specimen of living material of the Spirillum of cholera, addressed to those most able to grant it, was refused, and my other occupations precluded the possibility of my travelling in search of the disease. I still call Koch's form and others like it by the name of Spirillum, simply for the sake of shortness and simplicity. Hueppe proposes that it should be called Spirochaete ; Schröter would use the word Microspira as the generic name for the arthrosporous spiral Bacteria. My only objection to this is that changes and shifting of names in this group of organisms appear to me at present to be of little use and not desirable. The different species are still so unequally known that fresh changes may at any moment be required, and the best plan, therefore, is to be content with a simple intelligible expression for each case, and await the time when our knowledge will allow of our introducing a correct nomenclature, and one that may last for some time.

76. See the compilation by Judeich and Nitsche, Lehrb. d. mitteleurop. Forstinsectenkunde.—Pasteur, Études sur la maladie des vers-à-soie, Paris, 1870, with notices of the literature.—Frank R. Cheshire and W. Watson Cheyne, The Pathogenic History and History under Cultivation of a new Bacillus (B. alvei), the cause of a disease of the hive-bee, hitherto known as foul brood, in Journ. of Roy. Microsc. Society, ser. 2, V.—S. A. Forbes, Studies on the contagious diseases of insects, in Bull. of the Illinois State Laboratory of Nat. Hist., II (1886).—Also Metschnikoff in Virchow's Archiv, XCVI, p. 178.

77. J. H. Wakker, Onderzoek d. Ziekten van Hyacinthen, Harlem, 1883, 1884.—See also Bot. Centralblatt, XIV, p. 315.—T. J. Burrill, Anthrax of fruit trees, or the so-called fire-blight of pear trees, and twig-blight of apple trees, in Proceedings of American Association for the advancement of

Science, XXIX, 1880 ;—Id., Bacteria as a cause of disease in plants, in the American Naturalist, July, 1881.—J. C. Arthur, Pear Blight, in Annual Report of the New York Agricult. Experiment Station for 1884 and 1885 ;— Id. in Botanical Gazette, 1885;—Id. in American Naturalist, 1885.— E. Prillieux, Corrosion de grains de blé, &c., par les Bactéries, in Bull. Soc. Bot. de France, XXVI (1879), pp. 31, 167.—Reinke u. Berthold, Die Zersetzung d. Kartoffel durch Pilze, Berlin, 1879.—van Tieghem, Développement de l'Amylobacter dans les plantes à l'état de vie normal, in Bull. Soc. Bot. de France, XXXI (1884), p. 283.

INDEX.

o

THE END.

Reprint Publishing

FOR PEOPLE WHO GO FOR ORIGINALS.

This book is a facsimile reprint of the original edition. The term refers to the facsimile with an original in size and design exactly matching simulation as photographic or scanned reproduction.

Facsimile editions offer us the chance to join in the library of historical, cultural and scientific history of mankind, and to rediscover.

The books of the facsimile edition may have marks, notations and other marginalia and pages with errors contained in the original volume. These traces of the past refers to the historical journey that has covered the book.

ISBN 978-3-95940-091-6

Made in
Germany

www.reprintpublishing.com

www.ingramcontent.com/pod-product-compliance
Lightning Source LLC
Chambersburg PA
CBHW072307210326
41519CB00057B/3050